SUBSIDENCE MANAGEMENT HANDBOOK

NARESH C. SAXENA

Dean,
Planning and Resource Generation
&
Professor,
Indian School of Mines
Dhanbad – 826 004

SCIENTIFIC PUBLISHERS (INDIA)

P.O. Box 91 JODHPUR

Published by:
PAWAN KUMAR
Scientific Publishers (India)
5-A, New Pali Road, P.O. Box 91
JODHPUR - 342 001
Tel.: +91-291-433323
Fax : +91-291-512580
E-mail: artrans@tantramail.com
www.scientificpub.com

ISBN: 81-7233-317-X

Lasertype set : Rajesh Ojha
Printed in India

DEDICATION

The handbook is dedicated to
My wife Sudha for all the pains she
has been taking in bearing with me,
and her encouragement;

My daughters Gabboo, Babli and Kukki
for their love and affection;

My son-in-laws Shridhar and Pintoo
for their love and affection;

My grandchildren Minnu, Chittoo and Riya
who are my inspiration.

PREFACE

Subsidence, also termed as land subsidence, has been attracting attention of one and all due to its impacts not only on the surface but also on the sub-surface and underground properties. Over the globe, the subsidence movements have been found to have taken place due to various natural and manmade activities and the Man has been trying to find out the ways and means for minimizing the damages, prediction of subsidence movements, measurement of the movements, and finally the management of, the subsidence movements. Vast areas on the globe have experienced subsidence movements and related impacts and as of now many areas are experiencing these movements. The importance of the subsidence movements enhances with the importance of land areas as, with the growth of the population the per capita availability of the useful land is decreasing.

Among the various subsidence causing activities, the author considers the underground mining in sedimentary deposits and the withdrawal of water from the underground sources are as the most important and prominent activities in the Indian context. Underground mining of sedimentary deposits has been going on in the country for more than 225 years while extensive pumping of water from the underground sources is comparatively a recent phenomenon.

The subsidence know-how in respect of the underground mining of sedimentary deposits, though well-developed, has not been properly documented and the published literature on this subject does not make the operators self-reliant to take on the spot decisions, which, in the opinion of the author is necessary for efficient running of the industry. Therefore, in this handbook efforts have been made to present the subsidence management methodology in such a manner that in majority of the situations, the operating managers/engineers would be in a abled position to take decisions on their own and only in the situations involving complications the expert advice will be required. The handbook is also expected to be useful to the planners and the officials of the regulatory agencies, e.g., the Directorate General of Mines Safety and the Indian Bureau of Mines. It will also be of use to the faculty in the teaching institutions and will be an asset in the libraries.

In respect of the ground water withdrawal, the subsidence know-how in the country is very limited and there is a need to develop this as quickly as possible, as the dependence of the industries and the population on this source of water is increasing day-by-day and more-and-more number of tube-wells are being made wherever, ground water is available without considering the consequences. Therefore, an effort has been made by the author in this handbook to bring forth the importance of this subject for the future of the Nation as also to present some suggestions on the management of the subsidence in the areas having ground water withdrawal. The author anticipates that the handbook will initiate action towards developing the understanding of the subsidence phenomenon with respect to the withdrawal of water from the underground sources and the consequences of these movements, ultimately leading to the management of the activities such that the damages are minimum possible. This will also help in the development of the know-how for the prediction of the subsidence movements due to ground water withdrawal. The handbook will thus, be useful to the planners and managers of projects involving in utilization of water from the underground water sources. The handbook will also be useful to the students of civil and other related engineering disciplines.

This handbook presents the three chapters — the first chapter gives rise an overview of subsidence due to underground mining and ground water withdrawal — the second chapter deals exclusively with the management of subsidence due to underground mining of sedimentary deposits — the subsidence management aspects of ground water withdrawal are dealt in the third chapter.

The author has used the subsidence know-how in respect of the underground mining of sedimentary deposits in studying more than 100 cases related to the extraction of coal seams underneath and in the vicinity of surface and underground properties in the Indian coalfields. Most of the recommendations made in these studies have been implemented with the results that the coal mining industry has been able to extract more than 20 million tonne of coal from underneath and in the vicinity of the surface properties. In addition, the prediction of the subsidence movements and their impacts have also been done for a number of mines for the development of their Environmental Management Plans. Some of the case studies are presented briefly in the text.

The author is thankful to his friends and colleagues who helped him in the development of the subsidence know-how and who worked with him in monitoring of the subsidence movements in coal mining areas in the country. Special mention is being made for Dr. B. Kumar, Scientist, CMRI, Dhnabad and S/Shri R. Sinha, B.N. Pandey, S.D. Tiwary, S.N. Prasad, S.K. Singh and R.K. Sinha also of CMRI, Dhanbad. The author is also thankful to the critics who left no stone unturned in having their say. Thanks are also due to Shri Durga Das Chandra for helping in preparation of the text of the handbook on the computor and to Shri A. Bhattacharya for helping in preparation of the drawings for the handbook. The author is grateful to the friends who encouraged him to bring out this handbook for the benefit of the Indian community.

Dhanbad

Naresh C. Saxena

CONTENTS

Preface *v*

1. **SUBSIDENCE – AN OVERVIEW** **1-19**
   - A. Causes of Subsidence 1
   - B. Impacts of Subsidence Movements 5
   - C. Controlled Subsidence 16
   - D. Potential Areas in India 17
   - E. Dangers to Underground Workings 18
   - F. Concluding Remarks 19
2. **SUBSIDENCE MANAGEMENT IN UNDERGROUND MINING AREAS IN SEDIMENTARY DEPOSITS** **20-83**
   - Introduction 20
   - A. Collection, Compilation and Study of Details 21
   - B. Prediction of Subsidence Movements 25
     1. Critical width-depth ratio (CW) 27
     2. Non-effective width-depth ratio (NEW) 27
     3. Maximum possible subsidence (Smax) 29
     4. Maximum possible slope (Gmax) 31
     5. Maximum possible strains (Emax) 31
     6. Maximum subsidence (S) 32
     7. Maximum slope (G) and strains (E) 32
     8. Subsidence, slope and strain profiles 34
     9. Subsidence, slope and strain contours 41
     10. Volume of subsidence trough 43
     11. Location of cracks 43
     12. Changes in surface contours and drainage pattern due to subsidence 43
     13. Subsidence movements at the locations of surface properties 44

14. Profiles of Railway lines, canals and roads 44

C. Prediction of Impacts 44

1. Problems to underground workings 45

2. Impacts on properties 46

D. Subsidence Management 47

1. Optimizing subsidence movements 48

(i) Reducing extraction thickness 49

(ii) Optimizing panel size 49

(iii) Planning with stowing 49

(iv) Alternate stowing and caving and barrier-less mining 50

(v) Multi-seam extraction with caving 52

(vi) Extraction planning with non-effective width-depth ratio 52

(vii) Partial extraction of bord and pillar developed areas 53

2. Impact mitigation 53

(i) Location of the craks 53

(ii) Changes in surface contours 53

(iii) Study of the impacts of the subsidence movements 56

(iv) Minimizing dangers to the underground workings 58

3. Management of subsided land for various uses 61

(i) Developing surface water bodies 61

(ii) Developing forestry, plantation and grassland 62

(iii) Development of subsided land for agriculture 63

(iv) Development of subsided land for construction purposes 63

(v) Construction of Roads over subsided areas 64

4. Converting impacts into resources 65

(i) Surface water bodies 65

(ii) Underground water bodies 65

E. Monitoring of subsidence movements 68

1. Case Studies 75

I. Sudamdih shaft and incline mines 75

II. Salanpur Colliery 76

III. Jamadoba 6 & 7 pits Colliery 77

IV. Jitpur Colliery 77
V. RK7 & RK New Tech Inclines 77
VI. Naheriya Mine 77
2. Concluding Remarks 83

ANNEXURES **84-132**

Annexure I	:	Subsidence Impact Matrix	84
Annexure II	:	Controlling Subsidence Movements by Reducing Extraction Thickness	88
Annexure III	:	Controlling Subsidence Movements by Optimising Width-depth Ratio	90
Annexure IV	:	Controlling Subsidence Movements by Planning Extraction with Hydraulic Stowing	91
Annexure V	:	Controlling Subsidence Movements by Planning Extraction Without Barriers and by Alternate Stowing and Caving	92
Annexure VI	:	Controlling Subsidence Movements by Suitably Locating the Barriers	95
Annexure VII	:	Extraction Planning With NEW Concept	97
Annexure VIII	:	Determination of the Factor of Safety of the Barrier Pillars, and Hence Their Designing	97
Annexure IX	:	Computation of the Volume of Subsidence Trough	99
Annexure X	:	Details of Jitpur Colliery's Case Study	100
Annexure XI	:	Details of RK7 & RK New Tech Inclines' Case Study	104
Annexure XII	:	Details of Naheriya Project's Case Study	124

3. SUBSIDENCE MANAGEMENT IN AREAS HAVING GROUND WATER WITHDRAWAL WITH SPECIAL REFERENCE TO INDIAN SITUATION **133-141**

A. Preamble 133
B. Phenomenon 133
C. Impacts 137
D. Land Subsidence Management 138
1. Collection and compilation of the details 138
2. Prediction of land subsidence 139
3. Monitoring 139

4. Impact assessment 140
5. Management measures 141
6. Concluding Remarks 141

BIBLIOGRAPHY 142-205

CHAPTER - 1

SUBSIDENCE — AN OVERVIEW

— *A General Survey*

All the activities the human beings do are either for survival or development. The development of the human civilization cannot be imagined without the use of the fossil fuels and minerals as natural resources obtained from the Earth's crust. The man will continue to mine these resources for future development.

Exploitation of the fossil fuels and minerals in any form, i.e., solid, liquid or gas, disturbs the natural equilibrium of the rock mass causing readjustments resulting in vertical and horizontal displacements in the planes on the surface, sub-surface and underground levels. Similar movements are also caused by many other human and natural activities. These movements are generally termed as subsidence.

Subsidences due to various causes have wide ranging impacts of varying magnitude not only on the surface features and structures but also on the properties at sub-surface and underground levels.

A. CAUSES OF SUBSIDENCE

The causes of subsidence movements can be broadly divided into two categories based on genetic and activity-wise classification as shown in Tables 1.1 and 1.2. The genetic causes relate to the activities inside the earth's crust, i.e., endogenic, and on the earth's crust, i.e., exogenic. The activity-wise causes relate to natural and human activities.

It is evident from the causes outlined in Tables 1.1 and 1.2 that at any given place the nature and magnitude of the subsidence movements and their impacts depend on the subsidence causing activities taking place. It is also to be noted that the natural causes and activities are beyond human control, while the manmade activities can be regulated and controlled.

Table 1.1: Genetic classification of subsidence causing activities

Endogenic activities	Exogenic activities
• Folding	• Removal of supports
• Faulting	• Weakening of supports
• Continental drift	• Increase in loading
• Earthquakes	
• Volcanism	
• Earth fissures	

Table 1.2 : Subsidence causing activities

Natural activities	Manmade activities
• Solution of rocks and minerals	• Ground water withdrawal
• Drainage of soils	• Petroleum and natural gas withdrawal
• Drifting of soils	• Open excavations for mining and other purposes
• Sliding of rocks	• Underground excavations for mining and other purposes
• Frosting and defrosting	• Differential loading of soils
• Movements along earth fissures	
• Tectonic activities	
• Rodents	

Some examples of the subsidence locations due to various causes are given in Tables 1.3 to 1.5, which indicate that ground water withdrawal, oil and gas withdrawal, geothermal fluid withdrawal, hydro-compaction, earth fissures and other activities have caused extensive subsidences.

Table 1.3 : Land subsidences due to ground water withdrawal, Some examples

Sl. No	Place/Area	Country	References
1	Houston-Galveston, Texas	USA	Holzer, 1984, Holdahl, et al, 1991
2	Picacho Basin, Arizona	USA	Holzer, 1984
3	San Joaquin Valley, California	USA	Holzer, 1984
4	Nobi Plain	Japan	Sato, Daito & Ueshita, 1984
5	West Siberian-Middle Asian Segment	Russia	Vartanyan & Kolikou, 1984
6	Chuniespoort Group Dolomite and Chert of Wondertontein Valley	South Africa	Brackley, Rosewarne & Grady, 1984
7	Niigata Plain	Japan	Yoshida & Aoki, 1984
8	Central Valley, California	USA	Williamson & Prudic, 1984
9	Phoenix, Arizona	USA	Larson, 1984
10	Venice City	Italy	Gevon, et al, 1984 Carbognin & Gatto, 1984
11	Delta Area, River Po	Italy	Gambardella Mercusa, 1984
12	The Yamagata Basin	Japan	Matsuoka, 1984
13	Bangkok City	Thailand	Bergado, Balasubramanian & Apaipong, 1984 Phien-wij, et al, 1989
14	Taiyuan City	China	Yu-sun, 1984
15	Shanghai	China	He-yuan, 1984 Tie-Zhu & Jie-ddia, 1989
16	Tianjin	China	Qingzhi & Xioujun, 1984 Qingzhi, 1989
17	The Netherlands	The Netherlands	Rietveld, 1984
18	Stockholm	Sweden	Hansbo, 1989
19	Nagoya	Japan	Ohshima, Ueshita & Daito, 1989
20	Mexico City	Mexico	Rivera, et al, 1989 Rivera, et al, 1991
21	Hulla Swamp	Israel	Shaham & Levin, 1989
22	Thessaloniki	Greece	Andronopoulos, et al, 1991
23	Southwest Taiwan	Taiwan	Liao, et al, 1991
24	Kelantan	Malaysia	Haryono, 1995
25	Hucco Basin, El Paso, Texas	USA	Haywood, 1995
26	Southwestern Flanders	Belgium	Lebbe, 1995
27	Hanoi	Vietnam	Nguyen & Helm, 1995
28	Sagamigawa Alluvial Plain, Kanagawa	Japan	Yokoyama, et al, 1995
29	Jakarta	Indonesia	Yong, et al, 1995
30	Modena & Bologna & other places	Italy	Balestri & Villani, 1991 Dazzi, et al, 1984 Astori & Bezoari, 1989

Table 1.4 : Land subsidences due to oil and gas withdrawal, Some examples

Sl. No	Place/area	Country	References
1	Ravenna area	Italy	Bertoni, et al, 1995
2	Belridge Diatomite oil field, Karn County, Callifornia	USA	de Rouffignac & Bondor, 1995; Bondor & de Rouffignac, 1995
3	Ekofisk, North Sea	UK	Eiksund, et al, 1995; Wiborg & Jewhurst, 1986
4	Groningen gas field	The Netherlands	Hoefnagels, et al, 1995; Schoonbeck, 1976; Barends, 1989
5	Goose Creek oil field, Texas	USA	Pratt & Johnson, 1926
6	Bolivar Coast oil field	Venezuela	Van der Knaap & Van der Vlis, 1967
7	Wilmington oil field, Long Beach, California	USA	Gilluly & Grant, 1949
8	Huntington Beach oil field, California	USA	Martin & Serdengeeti, 1984
9	The Northern Part	The Netherlands	Pottingens & Haimson, 1991
10	Miyazaki Plain gas field, Kyushu Island	Japan	Esaki, et al, 1991
11	Ravenna gas field	Italy	Gambolati, et al, 1991
12	Western Venezuela gas fields	Venezuela	Murria, 1991; Mendoza & Murria, 1989
13	Po Delta	Italy	Gambolati, et al, 1984
14	Hokkaido, Chiba, Niigata & Okinawa	Japan	Nishida, et al, 1984

Table 1.5 : Some examples of land subsidences due to other activities

Sl. No	Place, Area & Country	Cause/Activity	References
1	The Netherlands, USA, Russia & Ireland	Organic soil subsidence	Whittaker & Reddish, 1989; Stephens, et al, 1984
2	Geothermal field at Wairakei, North Island, New Zealand	Geothermal fluid withdrawal	Stilwell, et al, 1975; Narasimhan & Goyal, 1984
3	Larderelli, Italy	Geothermal fluid withdrawal	
4	Cero Prieto, Mexico	Geothermal fluid withdrawal	
5	The Geysers, California, USA	Geothermal fluid withdrawal	

6	Great Britain	Underground coal gasification	Trent & Langland, 1985; Southerland, et al, 1983; Jegbefume & Thompson, 1983
7	Crater Lake, Oregon, USA	Volcanic activity	Prokopovich, 1984
8	Central Valley, California, USA	Folding	Prokopovich, 1984
9	Basin & Range Province, Nevada, USA	Faulting	Prokopovich, 1984
10	Phoenix, Arizona, USA	Land fissuring	Larson, 1984
11	Hula Swamp, Israel	Consolidation & compaction of soils	Levin & Shoham, 1984
12	Ijsselmeerpolder 'Flevoland', The Netherlands	Compaction of sediments	de Glopper, 1984
13	Northwest Germany, Germany	Drainage, evaporation & oxidation in peatland	Eggelsmann, 1984
14	New Orleans, USA	Drainage of soils	Snowden, 1984
15	San Joaquin Valley, California, USA	Hydrocompaction	Prokopovich, 1984
16	Gainesville, Florida, USA	Consolidation of clays	Ruth & Degner, 1984
17	Central Arizona Project, Arizona	Earth fissures	Sandoval & Bartlett, 1991
18	San Marcial & Deming, New Mexico, USA	Earth fissures	Heneberg, et al, 1991
19	Hudspeth County, Texas, USA	Earth fissures	Keaton & Shlemon, 1991
20	Temecula, Southwestern Riverside County, California, USA	Earth fissures	Carwin, et al, 1991
21	Mklimbres Basin, Southwestern New Mexico, USA	Earth fissures	Contaldo & Mueller, 1991
22	Sudbury Provincial Building, Canada	Pile driving	Peaker, et al, 1991
23	The North Sea, The French Atlantic Coast, Mediterranean, Australian Continent	Isostatic adjustment of the earth	Lambeck & Johnston, 1995

B. IMPACTS OF SUBSIDENCE MOVEMENTS

Subsidence movements of varying magnitude have caused almost negligible to severe impacts on surface, which depend upon the following:

* Nature, magnitude and extent of the subsidence movements.
* Nature of surface properties, i.e., agricultural area, built-up area, etc.

- Proximity of surface water bodies, i.e., sea, lakes, rivers, ponds, etc.
- Landscape and topography of the area.

The general characteristics of the impacts of subsidence movements are as outlined hereunder and given in Tables 1.6 to 1.10 with some examples.

Table 1.6 : General impacts of subsidence movements of the environment & surface properties

Impacts on environment	Impacts on surface properties
• Undesired depressions	• Damage to buildings and structures due to tilting, sink holes and development of curvature on the surface
• Uplifting of ground	• Damage to roadways, railway lines, etc. due to strains developing on the surface and due to sink holes
• Open cracks	• Breaking of pipe lines, sewage system, cables, etc.
• Water-logging and flooding	• Loss of land and properties due to inundation/flooding
• Disturbances in water regime	
• Air pollution and damage to land due to mine fires	
• Water pollution	
• Sink holes	

Table 1.7 : Damage to surface properties due to solid mining subsidence, Some examples

Sl. No	Colliery/Mine	Type of subsidence/damage
1	Ramnagore colliery, RCF	Four sudden subsidences over Begunia seam old workings damaging buildings and structures in Kendua Bazar and Kulti Steel Foundry
2	Khas Kajora colliery, RCF	Subsidence and sink holes (pot holes) in 1978, over the workings of Jambad and Kajora seams damaging a large number of quarters and bunglows
3	Girmint colliery, RCF	In 1993, a sudden subsidence, damaging a large number of quarters and private buildings
4	Mahabir colliery, RCF	Collapse and subsidence with sink holes due to drainage of water from old workings on two occasions
5	Sudamdih Project, JCF	A large scale subsidence in 1975 damaging a football field with sink holes over old workings

6	South Gobindpur colliery, JCF	A sudden subsidence causing damages to quarters and a pipe line
7	Digwadih colliery, JCF	Damage to miner's quarters and bunglows due to gradual subsidence
8	Sijua colliery, JCF	Damage to miner's quarters due to gradual subsidence
9	Moonidih Project, JCF	Planned damages to the buildings and quarters due to subsidences over caved underground workings
10	Simla Bahal colliery, JCF	Subsidence in 1982 due to collapses in XIV seam damaging houses in Simla Bahal Bustee
11	Alkusa colliery, JCF	Damage to miner's quarters in 1990 due to subsidence over XIV, XV & XVA seam workings
12	Kendwadih colliery, JCF	Two subsidences with sink holes in 1976 and 1981 damaging a railway siding and buildings
13	Mosabani copper mine, Bihar	Damage to some hutments in a village due to the subsidence in 1994
14	Khetri copper mining complex, Rajasthan	Subsidence causing a marked scar and a depression in a hill
15	Mochia mine in Zawar mining complex, Rajasthan	Subsidence causing a marked scar and a depression in a hill

RCF – Ranigunj coalfield JCF – Jharia coalfield

Table 1.8 : Controlled subsidences for the extraction of the solids below surface properties, Some Indian examples

Sl. No	Colliery/Mine	Extraction & Subsidence
1	Sudamdih Project, JCF	Extraction of VIIIA, IX/X and XI/XII seams below a main railway line and Damodar river with hydraulic sand stowing. The railway line was gradually made to subside by a maximum of about 800 mm.
2	Jitpur colliery, JCF	Extraction of XVI & XVIA seams below surface properties with caving and stowing. Extraction of XVII & XVIII seams with stowing below quarters, a railway siding, etc.
3	Jamadoba 1 & 2 Pits colliery, JCF	Extraction of XVA seam with caving using limited span method below built up area
4	Jamadoba 6 & 7 Pits colliery, JCF	Extraction of X and XII seams by longwall stowing below Bhaga railway station and yard
5	Salanpur colliery, JCF	Extraction of X seam below Katras railway yard
6	Loyabad colliery, JCF	Extraction of XIV seam below built up area. Some buildings were slightly damaged and then repaired.
7	Moonidih Project, JCF	Extraction of XVIIT, XVIT & XVIB seams by longwall caving below villages, railway siding, buildings, etc. Some buildings and quarters were damage and repaired.
8	Dhemomain colliery, RCF	Extraction of Disergarh seam by bord and pillar caving below build up areas by narrow panel method

9	Ningah colliery, RCF	Do -
10	Surakachhar colliery, Korba coalfield	Extraction of G III seam below surface properties and a railway siding with caving as well as stowing
11	Balgi Project, Korba coalfield	Extraction of GI seam below surface water bodies without stowing
12	Sripur colliery, RCF	Extraction of Koithee seam by bord and pillar caving underneath surface properties and Rana seam workings
13	Girmint colliery, RCF	Extraction by bord and pillar caving in Rana seam in the vicinity of a railway line.

Table 1.9 : Subsidences and related problems due to fluid withdrawal, Some examples

Sl. No.	Location	Activity	Subsidence & related problems
1	Ekofisk, North Sea (Maury, Sauzay & Fourmaintraux, 1987)	Oil exploitation	Tilting of oil tank and reduction of clearance from sea bed to platform
2	Central Valley, California, USA (Prudic & Williamson, 1984)	Ground water withdrawal	Decline in water level and subsidence due to large scale water withdrawal since early 1900s.
3	Nobi Plain, Japan (Sato, Diato & Ueshita, 1984)	- Do -	Subsidence in an area of about 1300 sq km since early 1980s, it was also correlated with earth quakes
4	Western Siberia, Tyan-shan-Kazakh & Turan region (Vartanyan & Kulikov, 1984)	- Do -	Land slide formation, water level rise at urban and agricultural territories
5	Modena, Italy (Russo, 1984)	Soil sinking	Damages to historical buildings in Modena city
6	Bologna, Italy (Pieri & Russo, 1984)	- Do -	Damage to old buildings in north Bologna area. Leaning of Asinelli tower.
7	Phoenix metropolitan area, Arizona, USA (Larson, 1984)	Ground water withdrawal	Opening of earth fissures and surface cracks, etc.
8	Venice, Italy (Carbognin & Gatto, 1984)	- Do -	Flooding during High Tide and damage to structures
9	Delta area of river Po, Italy (Gambardella & Mercusa, 1984)	Ground water & gas withdrawal	Damage to hydraulic system of the land
10	Tokyo & Osaka cities, & Yagamata basin, Japan (Matsuoka, 1984)	Ground water withdrawal	Drying of artesian wells and other problems
11	Bangkok, Thailand (Bargado, et al, 1984)	- Do -	Damage to side walk and buildings, etc.
12	Taiyun city, Shauxi province, China (Yu-sun, 1984)	- Do -	More than 1.2 m subsidence and resulting problems

13	Tianjin city, China (Qingzhi & Xioujun, 1984)	- Do -	More than 2.12 m subsidence and resulting problems including damage to the buildings
14	The Netherlands (Reitveld, 1984 & Suclting & Bruchem, 1984)	Coal mining, and withdrawal of gas and water	Damage to buildings, etc.
15	Hula Swamp, Israel (Levin & Shoham, 1984)	Drainage of soil	Lowering of ground water level and increase in frequency and extent of flooding
16	Metropolitan New Orleans, Louisiana, USA (Snowdon, 1984)	Drainage of reclaimed land	Structural failure of driveways, walkways and rupture of underground utility pipes
17	San Joaquin valley, California, USA (Prokopovich, 1984)	Hydrocompaction	Development of cracks, sink holes, and damage to buildings and structures, etc.
18	Groningen gas field, The Netherlands (Pottgens, 1984)	Gas exploitation	A budget of US$ 350 million was reserved for repair work and counter measures due to subsidence
19	Hokkaido, Chiba, Niigata, Miyazaki, Okinawa, etc., Japan (Nishida, et al, 1984)	- Do -	Damage to roads, bridges, houses, etc.
20	Houstan-Galveston, Texas, USA	Water and oil exploitation	Flooding of coastal areas
21	Sweden (Hansbo, 1989)	Ground water withdrawal	Water leakage into tunnels, damage to pavements, etc.
22	Costa,oriental oilfields, western Venezuela (Mendoza & Murria, 1989)	Oil, water and gas extraction	More than 4.6 m of subsidence causing it necessary to construct a dyke to protect urban areas and production facilities
23	Milan, Plain of Pisa, Como & Pistoia (Astori & Bezari, 1989)	Ground water withdrawal	Damages to buildings and structures
24	Nagoya, Japan (Ohshima, et al, 1989)	- Do -	Flooding from sea water
25	Valley of Mexico, Mexico	- Do -	Cracking of buildings, rupturing of conduits, etc.
26	Thessaloniki, Greece (Andronopoulos, et al, 1991)	- Do -	Sea intrusion
27	Linpien, Chiatung & Shuitiliao, Southwest Taiwan (Liao, et al, 1991)	- Do -	More than 2 m of subsidence and resulting problems
28	Northern Netherlands, The Netherlands (Pottgens, et al., 1991)	Gas extraction	Less production in farming and agriculture. Decrease in safety as a result of lowering of dykes and erosion of dunes
29	Miyazaki Plain in Kyushu Island, Japan (Esaki, et al., 1991)	Gas and ground water withdrawal	About 300 mm of subsidence between 1974 and 1988 and related problems

30	Ravenna, Italy (Gambolati, et al, 1991; Bertoni, et al, 1995)	- Do -	More than 1.3 m subsidence and related problems to urban infrastructures
31	Baytown, Texas, USA (Potok, 1991)	- Do -	Tidal and permanent flooding
32	Kanto Plain, Japan (Nakao & Kamata, 1991)	Ground water withdrawal	More than 1.0 m of subsidence between 1974 and 1989 and related problems
33	Owari Plain, Japan (Daito, et al, 1991)	- Do -	Large scale subsidence and related problems
34	Kelantan, Malaysia (Haryono, 1995)	- Do -	Not defined
35	Flanders, Belgium (Lebbe, 1995)	- Do -	- Do -
36	Hanoi, Vietnam (Nguyen & Helm, 1995)	- Do -	Damage to houses, schools and factories
37	Belridge Diatomite Oil Field, Kern County, California, USA (de Rouffignac & Bondor, 1995)	Oil exploitation	Well failures and large fissures due to about 3 m of subsidence

Table 1.10 : Problems caused due to sink hole subsidences – Some examples

Sl. No.	Location	Cause	Problem
1	Norwich, Norfolk, Great Britain (Price, 1995)	Chalk and flint mine's underground cavities/ openings' failure associated with heavy rain fall, blasting and earthquakes, etc.	Damage to buildings. Two fatalities in one case
2	Houd Berkaoui oil field, Algerian Sahara (Akretche & Okbi, 1995)	Formation of natural underground cavities	A 200 m diameter and 80 m deep crater and another 320 m diameter crater which swallowed one oil well each
3	US strategic pertoleum reserve storage site, Weeks Island, Louisiana, USA (Neal&Myers 1995)	Dissolution of salt below sink hole	Damage to a residential building
4	Saudi Arabia (Amim & Baukher, 1995)	Solution of host rocks in carbonate, salt diapers and evaporite terrains	Not specified
5	South Wales coalfield, UK (Statham & Treharne, 1991)	Opening failure, etc.	400 incidents. 2/3rd in open land, 16% caused damage to roads and structures and only one case of injury
6	West Midlands, UK (Brook, 1991)	Collapse of cavities/ openings in old limestone mines	Damage to industrial buildings

7	Ebro Basin, NE Spain (Benito & Del Campo, 1991)	Solution process in evaporites (gypsum & halite)	Cracks in buildings. Road subsidence & damage to irrigation channels
8	USA (Newton, 1984)	Collapse of natural and manmade cavities	6,000 sink holes at 850 sites in 19 states causing extensive damages.
9	Damascus Gate, Jerusalam, Isreal (Arkin, 1984)	Collapse of a cavity in limestone bed rock	8 m diameter and 6 m deep sink hole at a junction of roads
10	Cheshire, UK (Howell, 1984)	Collapse of cavities in rock salt beds	Severe effects in engineering structures, buildings, roads, railways, canals, etc.

- Development of marked depressions on the surface with or without open cracks (Fig. 1.1).
- Uplifting of ground and sometimes formation of ridges on the surface (Fig. 1.2).
- Water-logging in the depressions during rainy seasons and due to inundation from nearby water bodies if the surface areas subside below their highest flood level/highest tide level (HFL/HTL) (Fig. 1.3).
- Development of sink holes or pot holes (Fig. 1.4).
- Damage to buildings and structures due to tilting, curvature and strains, and development of sink holes (Fig. 1.5 & 1.6).
- Damage to roads, railway lines, etc. due to curvature and strains developing on the surface.
- Breaking of pipe-lines, sewage system, cables, etc.
- Loss of land and properties due to inundation.
- In some situations it may become necessary to shift and resettle the population affected due to subsidence.
- Loss of agricultural production from the subsided land.
- Marked decrease in water availability from underground sources, i.e., from aquicludes and aquifers.
- Development of mine fires and related environmental and societal impacts, specially in coal mining areas.
- In the plains after flooding the water from the subsided areas does not go back and remains stagnant in the depressions.

Fig. 1.1. Marked depressions due to subsidence

Fig. 1.2. Formation of ridges due to subsidence

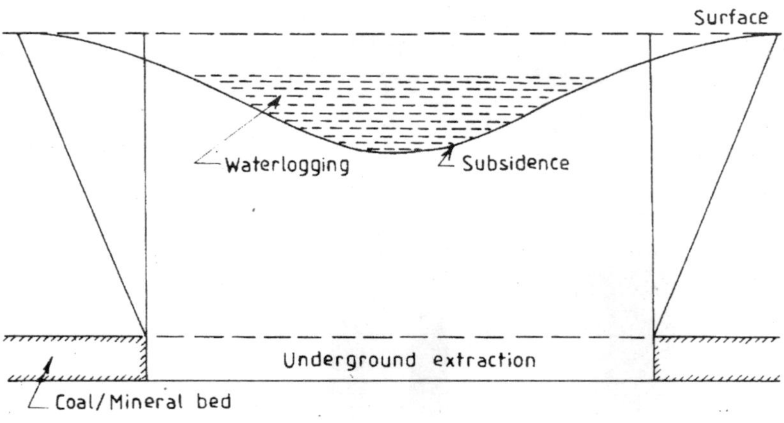

Fig. 1.3. Water-logging due to subsidence

Fig. 1.4. A pot-hole/sink-hole

Fig. 1.5. Damage to buildings due to subsidence

Fig. 1.6. Damage to buildings due to subsidence

C. CONTROLLED SUBSIDENCE

Subsidence movements due to manmade activities can be controlled to a large extent by suitable planning. In fact with the present status of technology it is possible to predict the subsidence movements due to the manmade activities and control their magnitude and nature. In addition to the manmade activities, efforts are being made to predict subsidence movements due to some natural activities and such predictions have already been made for the continental drift.

There are numerous examples of planned and controlled subsidences due to human activities, a few are mentioned below. Some Indian examples of planned coal mining subsidences are given in Table 1.8. A few case studies of the Indian coal mining areas are discussed in detail in the next chapter.

The Netherlands has regulated the rate of subsidence due to fluid withdrawal by controlling the rate of withdrawal and by taking other suitable measures.

At Houston, USA, the rate of ground water withdrawal was controlled to reduce the rate of ground subsidence.

An area in an inland harbour in Germany was made to gradually subside in a planned manner by underground extraction of coal seams.

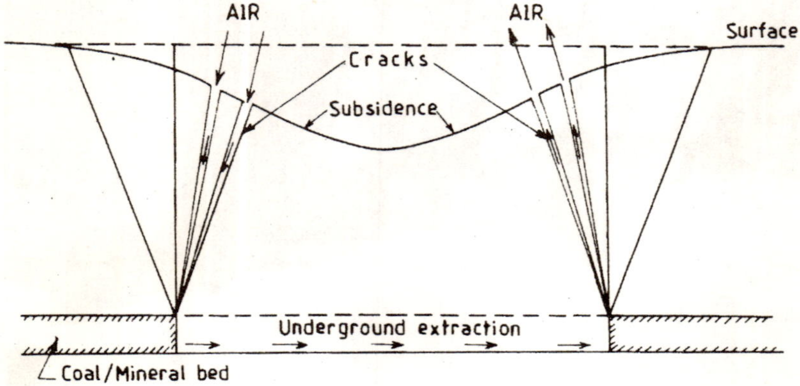

Fig. 1.7. Leakage of air from underground workings to the surface

D. POTENTIAL AREAS IN INDIA

There are many areas in India, which are either experiencing subsidence or are likely to experience subsidence due to human activities. These are outlined in Table 1.11. It is seen that apart from the coal mining areas, the activity of ground water withdrawal has a rather high subsidence potential and hence it needs immediate attention.

Table 1.11 : Areas likely to experience subsidence in India

S. No.	Activity	Areas likely to experience subsidence
1	Ground water withdrawal	• All cities and towns using ground water for industrial and domestic purposes, e.g., Kolkata, Delhi, Kanpur, Ahmedabad, Bangalore, Nagpur, Bareilly, Lucknow, Meerut, Chandigarh, etc.
		• Agricultural areas meeting water requirement from underground sources, e.g., plains and agricultural areas in Punjab, Harayana, Gujarat, Madhya Pradesh, Uttar Pradesh, Bihar, West Bengal, etc.
		• All opencast and underground sedimentary deposit mining areas
2	Withdrawal of petroleum and natural gas	• On-shore oil and gas fields in Gujrat and North Eastern States.
		• Off-shore oil and gas fields in Arabian sea, e.g., Bombay High and Bassien.
3	Mining areas	• Opencast and underground coal mining areas in the states of Assam, Meghalaya, West Bengal, Jharkhand, Orissa, Chattisgarh, Uttar Pradesh, Madhya Pradesh, Maharashtra and Andhra Pradesh.
		• Lignite mining areas in Tamil Nadu and Rajasthan.
		• Underground rock phosphate mining areas in Uttar Pradesh and Rajasthan.
		• Limestone mining areas in many states.

E. DANGERS TO UNDERGROUND WORKINGS

Besides the impacts on surface, sub-surface and underground properties the subsidence movements can cause some dangers to the underground mine workings. These are briefly outlined hereunder.

- In case of shallow underground mine workings the air being circulated in the underground workings for ventilation may find its way to the surface through the cracks formed due to subsidence (Fig. 1.7). This may result in poor ventilation and thus may cause an undesirable increase in the concentration of methane (CH_4) and carbon-monoxide (CO). In order to keep the concentration of these gases below the permissible limits greater quantity of air will be required for ventilation.
- Leakage of air through caved goaves, specially in bord and pillar depillaring, may cause spontaneous heating of coal left in the goaf, which may lead to underground mine fires and finally the surface fires. Many fires existing in the Indian coalfields started as a result of this phenomenon.
- Due to the development of the cracks, from the underground workings to the surface, the underground water bodies over the workings, i.e., the water table and the aquifers, get disturbed (Fig. 1.8). As a result, the water from such water bodies may find its way to the underground workings, thus increasing the make of water in the mine. In extreme conditions this may cause inundation.

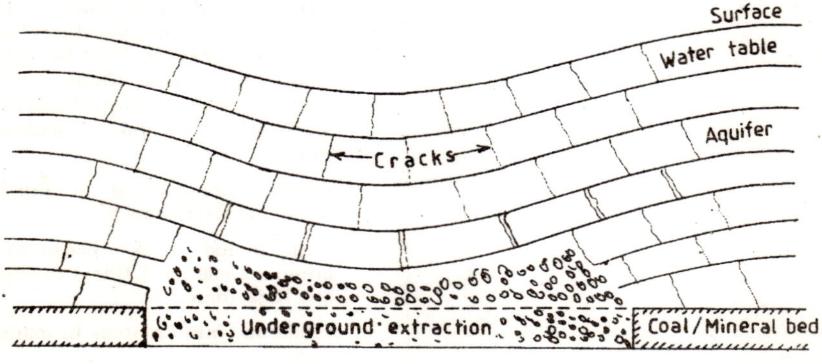

Fig. 1.8. Disturbances in underground water bodies due to subsidence

- The rain water may find way from the surface to the underground workings through cracks and if the ground has subsided to a level lower than the HFL/HTL of nearby water bodies, the mine may get inundated (Fig. 1.3).

It may be noted from the above that all the dangers to the underground workings are due to the development of cracks in the strata from the underground workings to the surface.

F. CONCLUDING REMARKS

It is evident from the foregoing observations that subsidence movements due to various activities cause impacts of varying degree and therefore it is necessary that subsidence management strategies are developed for the areas where the subsidence causing activities are taking place. Aspects of subsidence management discussed in the following Chapters are given below.

- Subsidence management in sedimentary deposits in underground mining areas.
- Subsidence management in areas having ground water withdrawal.

CHAPTER – 2

SUBSIDENCE MANAGEMENT IN UNDERGROUND MINING AREAS IN SEDIMENTARY DEPOSITS

INTRODUCTION

The following fossil fuel and sedimentary minerals are being mined in India by the underground mining methods.

- Fossil fuel : Coal
- Minerals : Rock phosphate and Pyrite

In near future with the increase in the environmental awareness and concern some more minerals, e.g., limestone, are likely to be mined by the underground mining methods.

As is evident from the discussions in the previous chapter underground mining of minerals in sedimentary deposits causes subsidence which not only affects the surface, sub-surface and underground properties but also the underground mine workings. Therefore, it is necessary that underground mine planning should take into consideration the aspect of ***subsidence management*** so as to

- afford continuous use of land on the surface, and
- minimize the dangers of subsidence movements to underground mine workings.

In fact, subsidence management should be an integral part of underground mine planning as this, besides the requirements under the Coal Mines Regulations 1957, will also take care of the requirements under the Environment (Protection) Act 1986 and the Forest (Conservation) Act 1980. In addition, the impacts on the local community will be minimum because subsidence management affords minimum loss of land-use and with suitable precautions can augment water availability and improve land-use.

The steps involved in subsidence management planning for underground mining in sedimentary deposits are as outlined below.

1. Collection, compilation and study of the details of the mineral bearing area and proposed mining.
2. Prediction of subsidence movements and their optimization.
3. Prediction of the impacts of the subsidence movements and their optimization.
4. Mitigation of impacts and management of the subsided areas.

The above aspects of the subsidence management in sedimentary deposit underground mining along with a few case studies are outlined in this chapter.

A. COLLECTION, COMPILATION AND STUDY OF DETAILS

The following details are required for the study of the area from the stand point of the subsidence management planning.

1. *Surface plan of the area showing*
   (i) Surface topography with 1.0 m contour intervals
   (ii) Land-use
   (iii) Locations of water bodies and their HFL/HTL and drainage pattern
   (iv) Locations of villages, bustees, roads, railway lines, pipe lines, transmission lines, aerial ropeways, etc.
   (v) Locations of opencast mines, if any, etc.
2. *Geological details*
   (i) Bore-hole sections with detailed lithology
   (ii) Strata sections
   (iii) Details of water table and aquifers
   (iv) Locations and nature of geological disturbances, e.g., faults, dykes, and sills

(v) Thickness and depth of coal/mineral bed

(vi) Any other relevant information

3. *Mining layout*

(i) Layout of panels to be extracted in each seam separately

(ii) Details of panels, i.e., length, width and height of extraction

(iii) Proposed method of extraction

(iv) Proposed extraction plan, i.e., period of extraction of panels and sequence

4. *Safe limits of subsidence movements for various categories of surface properties.*

The above details are required for mine planning and hence should normally be available with the planners. However, if some of the details are not available these may be obtained from relevant sources, e.g.,

- remotely sensed data from the National Remote Sensing Agency (NRSA),
- land use data from the National Bureau of Soil Survey and Land Use Planning (NBSSLUP),
- geological data from the Geological Survey of India (GSI),
- hydrological data from the Central Ground Water Board (CGWB), etc.

Usually 1.0 m contour interval topographical plans are not readily available and therefore such plans are required to be developed using the above details. With the help of the details collected the following plans and data are compiled.

1. Contour plan with 1.0 m contour interval, drainage pattern and HFL/HTL of surface water bodies.
2. Surface plan showing land-use and locations of surface features and structures.
3. Layout plans of individual seams overlain by surface contour plan and surface features and structures separately.
4. Both the types of plans under serial number 3 should also show the locations of the geological disturbances.
5. Strata sections showing macro geological structure over the panels proposed for extraction.

6. Characteristics of water table and aquifers above the panels and, if possible, their water flow details.
7. Details of the panels in a tabulated form (Table 2.1 gives the suggested format).

Table 2.1: Details of the panels

Sl. No.	Seam	Panel	Extraction thickness (m')	Depth variation (h)	Size of panel (wxL)	Proposed period of extraction	% of extraction from panel	Width-Depth ratio*	Rock mass factor (M)	Remarks
1	2	3	4	5	6	7	8	9	10	11

* with respect to minimum depth.

Note: In case of semi-steep and steep seams, the gradient of the seam(s) is to be included in this table.

The value of the rock mass factor 'M' is compiled with the use of the details in the strata section and respective position of the seams as outlined in Table 2.2.

Table 2.2 : Rock mass factor

Sl. No.	Nature of Strata	Rock mass factor 'M'
1	Totally disturbed strata[1]	1.0
2	Partially disturbed strata with the thickness of intact strata being less than five times the seam thickness under extraction	0.9
3	Partially disturbed strata with the thickness of intact strata being more than five times the seam thickness under extraction	0.8
4	Naturally weak and disturbed strata[2]	0.7
5	Strata having interbedded layers of sandstones and shales	0.6
6	Strata having predominantly sandstone beds	0.5

1 - As in case of thick seam mining by descending slicing when the top slice has been extracted with caving and the next slice is under consideration.
2 - Strata having predominantly clays, shales, mud-stones, etc.

The compiled details are supplemented by the permissible limits of subsidence, which in the Indian context are as given in Table 2.3.

Table 2.3 : Permissible limits of subsidence movements

Sl No.	Surface property	Permissible limits
1	Railway lines of jointed construction	Maximum strain = 3.0 mm/m Limiting operating gradient i. for passenger lines = 1 in 120 ii. for sidings = 1 in 100
2	Railway lines of welded construction	No subsidence movements
3	Surface water bodies	Maximum tensile strain = 3.0-4.5 mm/m
4	High tension pylons and pylons of aerial ropeways	Maximum strain = 3.0 mm/m Slope not to cause the displacement of the top-most point by not more than 1/3rd of the radius of the base
5	Buildings (pucca)	Maximum total compression or elongation of 60 mm for repairable slight damages
6	Forest land	Maximum tensile strain = 20 mm/m with the width of cracks not more than 300 mm

The compiled details are studied to assess the following for the purpose of the subsidence management planning.

1. Identification of the locations of the surface properties, etc. which are not to be subjected to practically any subsidence movement.
2. Identification of the locations of the surface properties, etc. above and adjacent to the panels in the seams proposed for extraction and the permissible limits of the subsidence movements for the properties.
3. Nature of the surface drainage pattern and the proximity of HFL/HTL to the panels.
4. Assessment of rock mass factor 'M' for all the panels using the strata section and proposed sequence to extraction.
5. Implications of the presence of geological disturbances, specially major faults and sills. The faults tend to arrest and also extend the subsidence trough as shown in Fig. 2.1. The sills, depending upon the thickness and the nature of rock mass tend to bridge the strata and hence delay the subsidence and in situations having very thick sills the subsidence movements may be arrested.
6. Likely consequences of disturbances in the water table and strata in terms of the increase in the make of water in the underground mine workings.

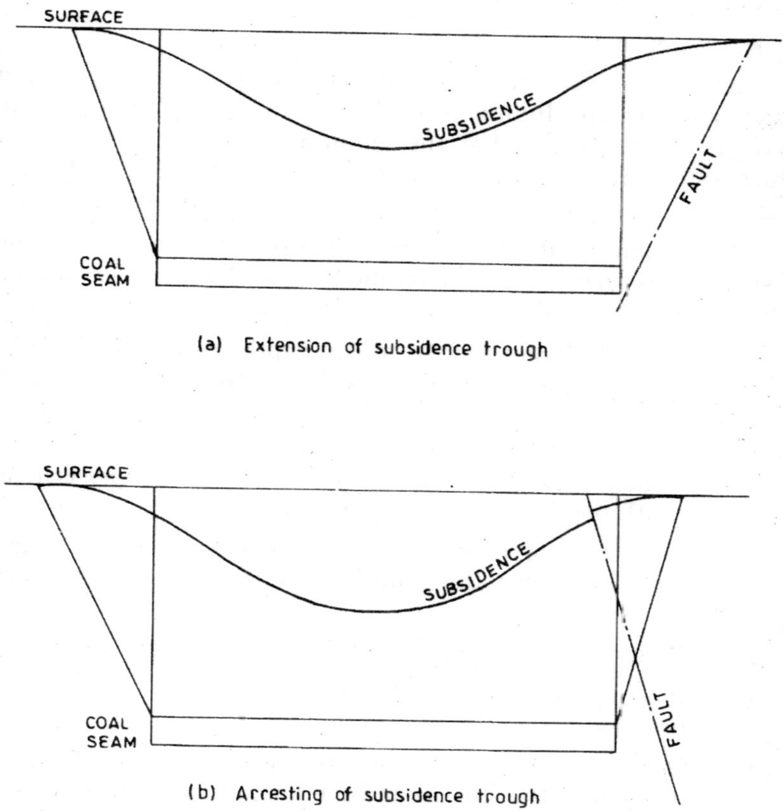

Fig. 2.1. Impacts of faults on subsidence trough

The above assessments are reviewed and if felt necessary the layout of the panels is modified so as to take care of the surface properties not to be subjected to any subsidence movement, interaction of faults with subsidence, etc. The layout of the panels thus developed is used for the prediction of subsidence movements.

B. PREDICTION OF SUBSIDENCE MOVEMENTS

There are numerous methodologies which can be used for the prediction of subsidence movements and the selection of the methodology is an important step. The methodologies can be classified into the following three categories.

- Mathematical/Numerical modeling
- Physical modeling
- Empirical modeling

All these have their limitations, scope, advantages and disadvantages. The Indian scenario analyzed by Chatterjee and Saxena, 1996 revealed that the empirical models/methodologies which satisfy the boundary conditions are most likely to be the useful ones. The boundary conditions, etc. are discussed later.

Prediction of subsidence movements either for the protection of surface properties, the environmental clearance or for mining below forest areas is done in the following sequence.

1. Defining non-effective width-depth ratio (NEW) and critical width-depth ratio (CW) for the different mining conditions.
2. Prediction of maximum possible subsidence (Smax), slope (Gmax) and strains (Emax) for critical and super critical width-depth ratios of extraction.
3. Prediction of maximum subsidence (S), slope (G) and strains (E) for the panels of sub-critical width-depth ratios.
4. Prediction of the volume of subsidence trough (V).
5. Plotting of anticipated subsidence, slope and strain profiles.
6. Defining the locations where the cracks are likely to develop on the surface and the magnitude of the cracks.
7. Outlining the changes in surface contours likely to occur due to subsidence.
8. Modeling the changes in surface topography in three-dimensional views due to subsidence.
9. Anticipated changes in drainage pattern on the surface.
10. Subsidence movements at the locations of surface properties.
11. In case of railway lines and roads it is required to predict the changes in their profiles.

The following or any other suitable methodology can be used for the prediction of the above parameters in relation to the subsidence movements.

1. Critical Width-Depth Ratio (CW) of extraction is the one which causes the occurrence of the maximum possible subsidence (Smax) in a given condition as shown in Fig. 2.2. Any further increase in the width-depth ratio does not increase the maximum subsidence and all the width-depth ratios more than the CW are known as super-critical width-depth ratios. The CW can be predicted, amongst others, by using equation (2.1) and Table 2.4.

$$CW = 2 - M \qquad (2.1)$$

Where CW = critical width-depth ratio, and

M = rock mass factor.

Table 2.4 : Critical width-depth ratio

Rock Mass Factor 'M' (Table 2.2)	1.0	0.9	0.8	0.7	0.6	0.5
Critical Width-Depth Ratio 'CW'	1.0	1.1	1.2	1.3	1.4	1.5

The panels having width-depth ratios more than the CW will also cause the maximum possible subsidence movements on the surface.

2. Non-Effective Width-Depth Ratios (NEW) of extraction do not cause practically any subsidence movement on the surface or at the plane of reference above the panels. The importance of this parameter lies in the fact that the extraction below the surface properties not to be subjected to practically any subsidence can be planned on the basis of this concept. Similarly in the situations where the parting is sufficient extraction in lower seam can be planned. The NEW concept has been used extensively all over the world for underground mine planning. Some examples from Indian coalfields are given in Table 2.5. The NEW for a given strata condition can be predicted, among others, by using equation (2.2) and Table 2.6.

$$NEW = 1 - M \qquad (2.2)$$

Where NEW = non-effective width-depth ratio, and

M = rock mass factor.

While planning the underground workings on NEW concept it would be necessary to leave barriers between the planned extractions. These barriers should be designed for long-term stability with a factor of safety more than 2.0.

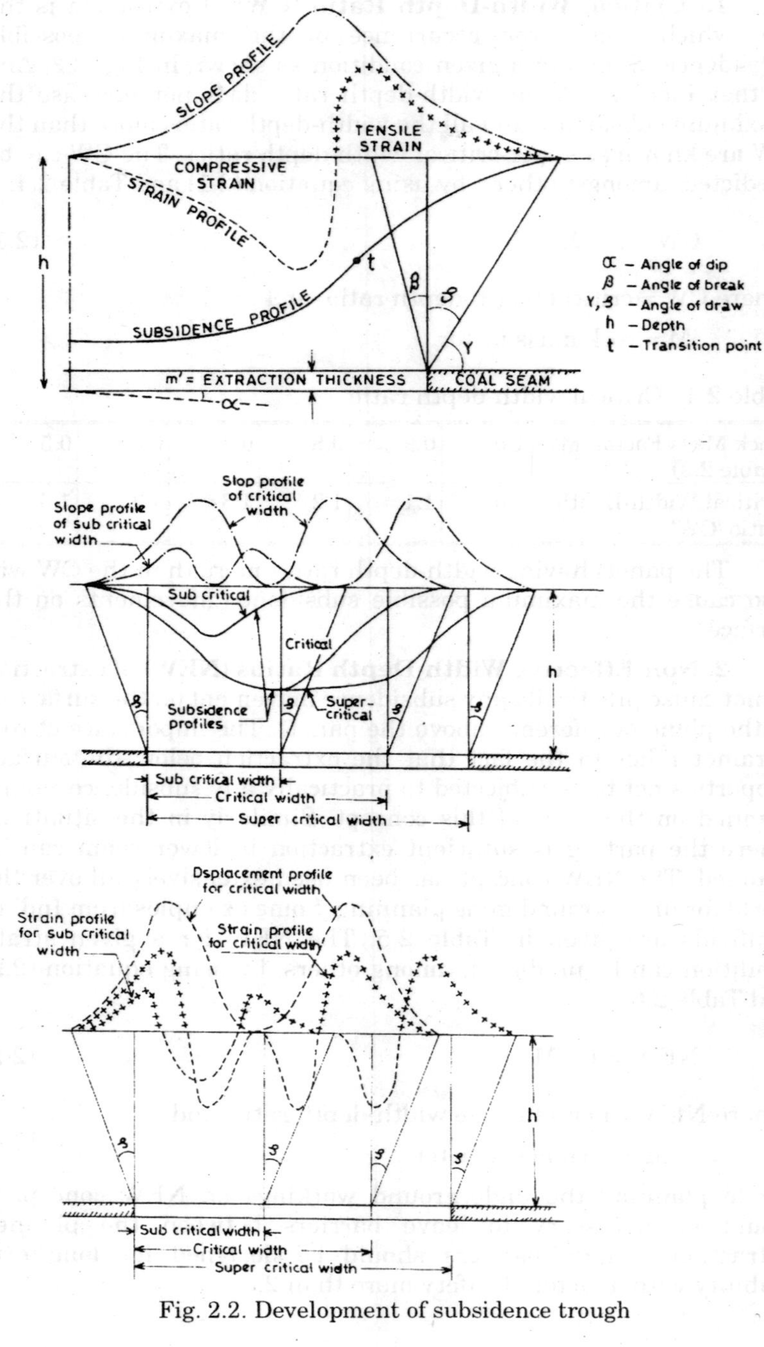

Fig. 2.2. Development of subsidence trough

Table 2.5 : NEW concept planning, Some examples

Sl No.	Mine	Seam	Condition
1	Sudamdih Incline Mine	XI/XII	Extraction below waterlogged workings in overlying XIV seam. The parting between the seams was about 100m
2	Jamadoba	XIV	Extraction below surface properties. The width of extraction was restricted to 0.3 times the depth.
3	Ningah	Disergarh	Extraction below surface properties by bord and pillar system with caving.
4	Moonidih	XVIII Top	Extraction below a large village by longwall system with caving.
5	Balgi	G I	Extraction below waterlogged G III seam workings and surface water bodies

Table 2.6 : Non-effective width-depth ratio

Rock Mass Factor 'M' (Table 2.2)	1.0	0.9	0.8	0.7	0.6	0.5
Non-Effective Width-Depth Ratio (NEW)	0.0	0.1	0.2	0.3	0.4	0.5

3. Maximum Possible Subsidence (Smax) (Fig.2.2) for any given geo-mining condition can be predicted, among others, with the use of equation (2.3), Table 2.7 and nomogram in Fig.2.3.

$$Smax = 500\,(1+M)\,(e'.\ a'.\ h'.\ m') \tag{2.3}$$

Where Smax = maximum possible subsidence in mm,

M = rock mass factor (Table 2.2),

e' = extraction percentage factor (Table 2.7),

a' = goaf treatment factor (Table 2.7),

h' = depth factor (Table 2.7), and

m' = extraction thickness in m.

Table 2.7 : Parameters in equation 2.3

Parameter		Value						
e′	Percentage of extraction	40	50	60	70	80	90	100
	e′	0	0.1	0.25	0.40	0.60	0.80	0.95
a′	Goaf treatment	Caving				Hydraulic stowing		
	a′	0.95				0.07-0.10*		
h′	Depth	<2.50 m		251-400 m		>400 m		
	h′	1.00		1.10		1.15		

* 0.07 for seams steeper than 1 in 5.

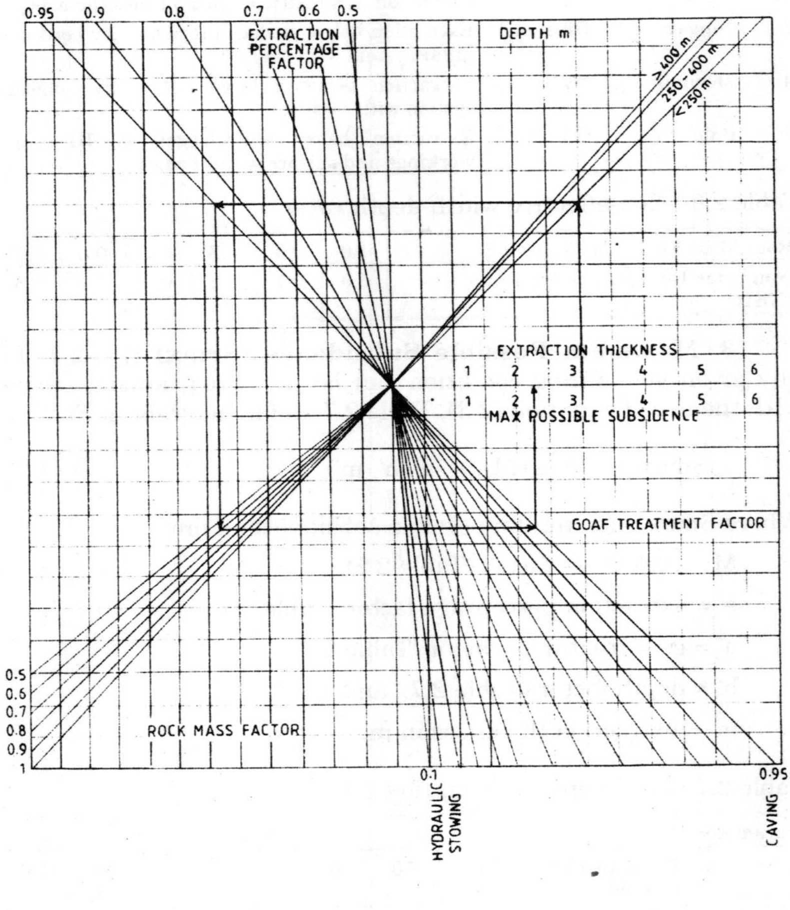

Fig. 2.3. Nomogram for predicting maximum possible subsidence

4. Maximum Possible Slope (Gmax) (Fig.2.2) can be predicted with the use of equation (2.4).

$$Gmax = 2.0 \frac{Smax}{h} \tag{2.4}$$

Where Gmax = maximum possible slope in mm/m,

Smax = maximum possible subsidence in mm, and

h = depth of the panel in m.

5. Maximum Possible Strains (Emax) (Fig. 2.2) can be predicted with the use of equations (2.5) and (2.6).

$$Emax_{(-)} = 1.0 \frac{Smax}{h} \quad \text{(compressive)} \tag{2.5}$$

$$Emax_{(+)} = 1.05 \frac{Smax}{h} \quad \text{(tensile)} \tag{2.6}$$

Where $Emax_{(-)}$ = maximum possible compressive strain in mm/m,

$Emax_{(+)}$ = maximum possible tensile strain in mm/m,

Smax = maximum possible subsidence in mm, and

h = depth of the panel in m.

EXAMPLE 1: Predict the CW, NEW and maximum possible subsidence movements for a 3.5 m thick stratified deposit being mined by (1) longwall system and (2) bord and pillar system with 80% possible extraction with caving at a depth of 200-230 m, the overlying strata being interbedded sandstones and shales.

PREDICTION:

(i) Rock mass factor = 0.6, as the strata over the deposit consists of interbedded sandstones and shales.

(ii) Critical width-depth ratio = 2.0 - 0.6 = 1.4

(iii) Non-effective width-depth ratio = 1.0 - 0.6 = 0.4

(iv) For computing Smax: M = 0.6; e' = 0.95 and 0.60; a' = 0.95; h' = 1.0; and m'= 3.5 m.

(v) Hence, Smax = 500(1+0.6)(0.95x0.95x1.0x3.5) = 2527 mm

$$Gmax = 2 \frac{2527}{200} = 25.27 \text{ or } 25.3 \text{ mm/m}$$

$$Emax_{(-)} = 1.0 \frac{2527}{200} = 12.6 \text{ mm/m}$$

$$Emax_{(+)} = 1.05 \frac{2527}{200} = 13.2 \text{ mm/m}$$

In this example the minimum depth has been taken into consideration as the maximum possible subsidence movements will be on the shallow side. These calculations are for condition (1) of extraction, i.e., for extraction by longwall system. For the second condition, i.e., for bord and pillar mining with 80% extraction the calculations will be as given below.

(i) For computing Smax: M=0.6; e′=0.60; a′=0.95; h′=1.0; m′=3.5 m.

(ii) Hence, Smax = 500(1+0.6)(0.60x0.95x1.0x3.5) = 1596 mm

$$Gmax = 2\,\frac{1596}{200} = 15.96 \text{ or } 16.0 \text{ mm/m}$$

$$Emax_{(-)} = 1.0\,\frac{1596}{200} = 7.98 \text{ or } 8.0 \text{ mm/m}$$

$$Emax_{(+)} = 1.05\,\frac{1596}{200} = 8.4 \text{ mm/m}$$

6. Maximum Subsidence (S) for any sub-critical width-depth ratio of extraction (Fig.2.2) can be predicted with the help of numerous models, equations, nomograms, etc. available in the literature on mine subsidence. For approximate and quick prediction the data given in Table 2.8 can be used.

Table 2.8 : Ready reckoner for prediction of maximum subsidence, slope and strains

Sl. No.	Rock Mass Factor 'M'						S/ Smax	K_1	K_2	K_3
	0.5	0.6	0.7	0.8	0.9	1.0				
	Width-Depth Ratio									
1	0.5	0.4	0.3	0.2	0.1	0.0	0.00	–	–	–
2	0.6	0.5	0.4	0.3	0.2	0.1	0.05	5.00	2.50	2.63
3	0.7	0.6	0.5	0.4	0.3	0.2	0.15	4.00	2.00	2.10
4	0.8	0.7	0.6	0.5	0.4	0.3	0.30	3.50	1.75	1.84
5	0.9	0.8	0.7	0.6	0.5	0.4	0.45	3.00	1.50	1.58
6	1.0	0.9	0.8	0.7	0.6	0.5	0.60	2.75	1.38	1.45
7	1.1	1.0	0.9	0.8	0.7	0.6	0.75	2.50	1.25	1.31
8	1.2	1.1	1.0	0.9	0.8	0.7	0.85	2.25	1.13	1.19
9	1.3	1.2	1.1	1.0	0.9	0.8	0.90	2.10	1.05	1.10
10	1.4	1.3	1.2	1.1	1.0	0.9	0.95	2.00	1.00	1.05
11	1.5	1.4	1.3	1.2	1.1	1.0	1.00	2.00	1.00	1.05

7. Maximum Slope (G) and Strains(E) for any given sub-critical width-depth ratio of extraction can be predicted with the help of numerous models, equations, and nomograms available in the literature. For approximate quick prediction the following equations and the data in Table 2.8 can be used.

$$G = K_1 \frac{S}{h} \tag{2.7}$$

$$E_{(-)} = K_2 \frac{S}{h} \text{ (Compressive)} \tag{2.8}$$

$$E_{(+)} = K_3 \frac{S}{h} \text{ (Tensile)} \tag{2.9}$$

Where, G = maximum slope in mm/m,

$E_{(-)}$ = maximum compressive strain in mm/m,

$E_{(+)}$ = maximum tensile strain in mm/m,

K_1, K_2 & K_3 = constants,

S = maximum subsidence in mm, and

h = minimum depth of panel in m.

The predicted values of maximum subsidence movements for each panel are presented in a tabulated form, for which an example is given in Table 2.9.

Table 2.9 : Maximum predicted subsidence movements

Sl. No.	Seam	Panel	Extraction thickness	Depth variation	Width-Depth ratio	Rock-mass factor	Maximum subsidence	Maximum slope	Maximum strains	
									Comp.	Tensile
			m	m			mm	mm/m	mm/m	mm/m
1	2	3	4	5	6	7	8	9	10	11

Before proceeding further it would be appropriate to discuss the above predictions with the help of an example, which is outlined below.

EXAMPLE 2: At a coal mine it is proposed to extract four panels each in two seams in superimposition by bord and pillar system with caving with 80% extraction. The strata overlying the upper seam is predominantly sandstone. The thickness of upper seam is 2.0 m and that of the lower seam is 3.0 m and the parting between the seams is about 30 m. Other details are as given in Table 2.10. Predict maximum subsidence movements due to extraction of the panels.

From the data given above the subsidence related parameters of the panels are as given below.

1. Upper seam : $m' = 2.0$ m; $M = 0.5$; $e' = 0.6$; $a' = 0.95$; $h' = 1.0$
2. Lower seam : $m' = 3.0$ m; $M = 0.8$; $e' = 0.6$; $a' = 0.95$; $h' = 1.1$

Table 2.10 : Details of the panels

Sl. No	Seam	Panel	Extraction thickness	Depth	Size of panel	Proposed period of extraction	% of extraction	Width-Depth ratio	Rock mass factor
			m	m	mxm		%		
1	2	3	4	5	6	7	8	9	10
1	Upper	1	2.0	225-230	130x500	–	80	0.58	0.5
2	Upper	2	2.0	230-235	170x500	–	80	0.74	0.5
3	Upper	3	2.0	235-240	190x500	–	80	0.81	0.5
4	Upper	4	2.0	240-245	240x500	–	80	1.00	0.5
5	Lower	1	3.0	255-260	130x500	–	80	0.51	0.8
6	Lower	2	3.0	260-265	170x500	–	80	0.65	0.8
7	Lower	3	3.0	265-270	190x500	–	80	0.72	0.8
8	Lower	4	3.0	270-275	240x500	–	80	0.89	0.8

The maximum possible subsidence (Smax) for the conditions in the two seams will be as given below.

1. Upper seam : Smax = 500 (1+0.5) (0.6x0.95x1.0x2.0)=885 mm

2. Lower seam : Smax = 500 (1+0.8) (0.6x0.95x1.1x3.0)=1692.9 mm

Using the data in Table 2.10 and the ready reckoner the predicted values of maximum subsidence movements are as given in Table 2.11.

Table 2.11 : Predicted maximum subsidence movements

Sl. No.	Seam	Panel	Extraction thickness	Depth*	Width-Depth ratio	Rock mass factor	Max. subsidence	Max. slope	Max. strain Comp.	Max. strain Tensil
			m	m			mm	mm/m	mm/m	mm/m
1	2	3	4	5	6	7	8	9	10	11
1	Upper	1	2.0	225	0.58	0.5	34	0.1	0.05	0.05
2	Upper	2	2.0	230	0.74	0.5	180	2.9	1.45	1.52
3	Upper	3	2.0	235	0.81	0.5	269	4.0	2.00	2.10
4	Upper	4	2.0	240	1.00	0.5	513	5.9	2.95	3.10
5	Lower	1	3.0	255	0.51	0.8	533	7.2	3.60	3.78
6	Lower	2	3.0	260	0.65	0.8	889	9.8	4.90	5.15
7	Lower	3	3.0	265	0.72	0.8	1067	10.9	5.45	5.72
8	Lower	4	3.0	270	0.89	0.8	1422	11.2	5.60	5.88

*Minimum depth

8. Subsidence, Slope and Strain Profiles (Fig. 2.2) : Numerous equations, models, etc. are available in the literature for predicting subsidence, slope and strain profiles due to underground mining. It is being suggested that only those equations which satisfy the boundary conditions should be used for prediction.

Boundary conditions

(a) For sub-critical width-depth ratios with origin at the center of subsidence trough (Fig. 2.4).

at $x = 0$, $s = S$, $g = 0$, $1/\rho$ = some value, and

at $x = \pm l$, $s = 0$, $g = 0$, $1/\rho = 0$.

(b) For sub-critical width-depth ratios with origin at the point of inflexion (Fig. 2.5)

at $x = 0$, $s = S/2$, $g = G$, $1/\rho = 0$, and

at $x = \pm l'$, $s = 0$ & S, $g = 0$, $1/\rho = 0$ and some value.

(c) For critical width-depth ratios with origin at the center of subsidence trough (Fig. 2.6).

at $x = 0$, $s = Smax$, $g = 0$, $1/\rho = 0$, and

at $x = \pm l$, $s = 0$, $g = 0$, $1/\rho = 0$.

(d) For critical width-depth ratios with origin at the point of inflexion (Fig. 2.7).

at $x = 0$, $s = Smax/2$, $g = Gmax$, $1/\rho = 0$, and

at $x = \pm l'$, $s = 0$ & $Smax$, $g = 0$, $1/\rho = 0$.

Where, x = distance of any point from the origin in m,

s = subsidence at the point x in mm,

g = slope of the subsidence profile in mm/m,

$1/\rho$ = curvature of the subsidence profile in m^{-1}, which is directly proportional to strains,

Smax = maximum possible subsidence in mm,

Gmax = maximum possible slope in mm/m,

S = maximum subsidence in mm,

G = maximum slope in mm/m,

$l = ½[W+2\,h \tan \gamma]$,

$l' = ¼\,[W+2\,h \tan \gamma]$,

h = depth of the panel in m,

W = width of the panel in m, and

s = angle of draw.

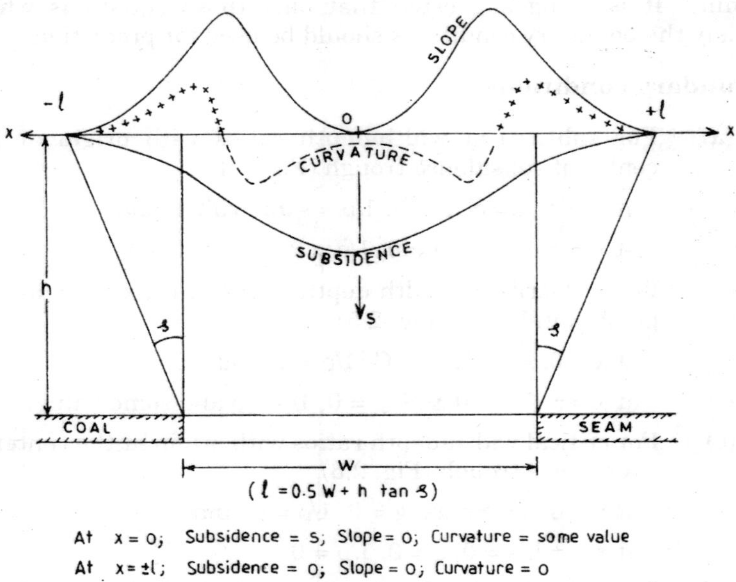

Fig. 2.4. Boundary conditions for sub-critical width-depth ratios with origin at the centre of the subsidence trough

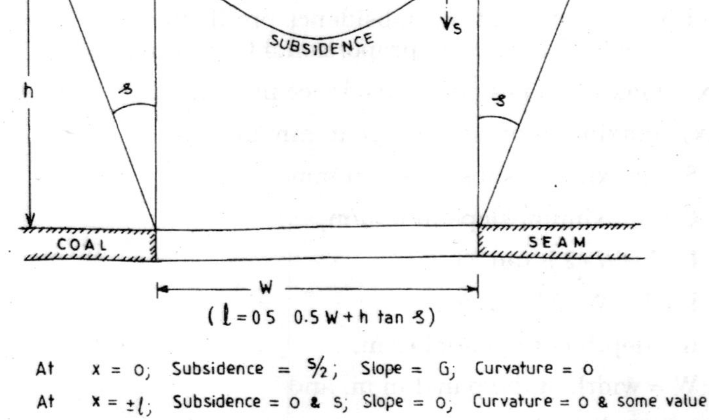

Fig. 2.5. Boundary conditions for sub-critical width-depth ratio with origin at the point of inflexion.

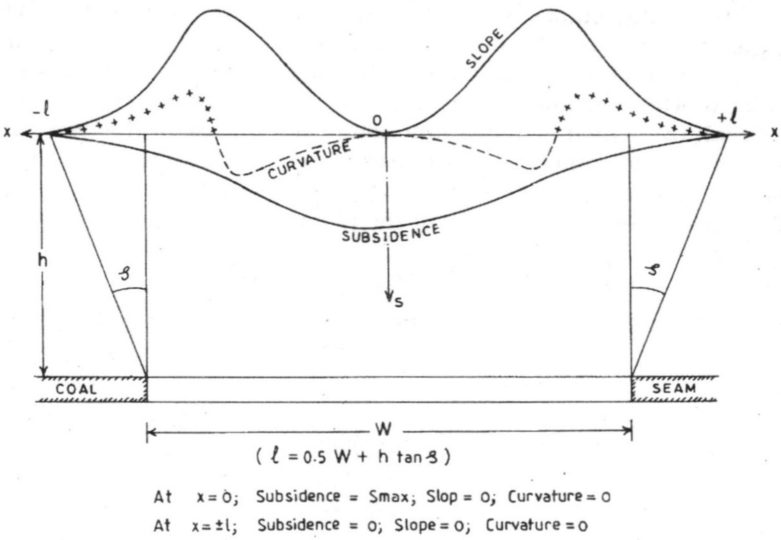

Fig. 2.6. Boundary conditions for sub-critical width-depth ratios with origin at the centre of the subsidence trough

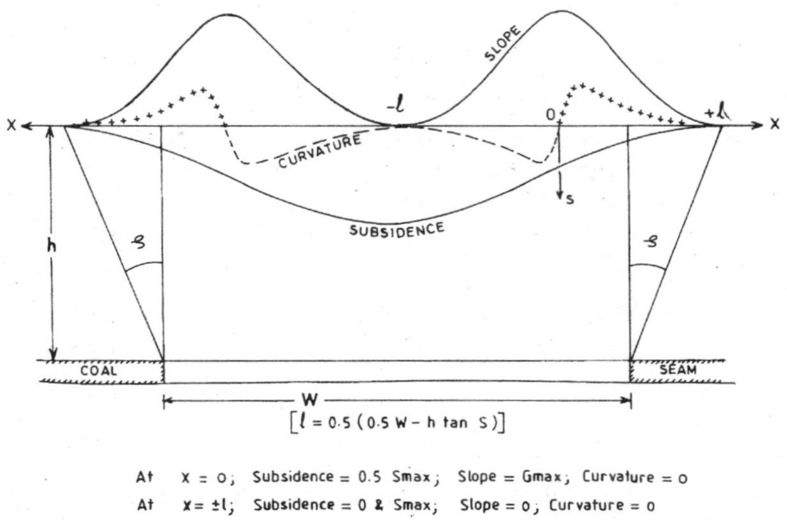

Fig. 2.7. Boundary conditions for sub-critical width-depth ratio with origin at the point of inflexion.

It is evident that for the assessment of the boundary conditions the knowledge of the angle of draw is essential. The angle of draw depends on the nature of the overlying rock mass and a rough prediction can be made with the use of data given in Table 2.12. Another parameter to be taken into consideration is the angle of break (ß), which is also a function of the nature of the overlying rock mass and can be approximately predicted with the help of the data in Table 2.12. The angle of break is always between the transition point and the edge of the panel.

Table 2.12 : Angle of draw and angle of break as shown in Fig. 2.2

Sl. No.	Rock mass factor(M)	Angle of draw(s)	Angle of break (ß)
1	1.0	40-45°	0°
2	0.9	38-42°	0-3°
3	0.8	35-40°	0-5°
4	0.7	30-35°	3-6°
5	0.6	20-25°	5-8°
0	0.5	15-18°	6-10°

In case at the time of prediction equations, models, etc. are not available the following methodology can be adopted for predicting the subsidence movement profiles.

(i) Obtain the value of the width (W) of the panel and its depth (h).

(ii) Assess the values of s and ß using Table 2.12.

(iii) Calculate the length of the subsidence trough (2l).

(iv) Plot on a graph paper the surface, the position of the underground panel, the angle of draw and the angle of break as shown in Fig. 2.8.

(v) Subsidence profile can be plotted by taking the following values.

- Maximum subsidence (S) at the central point,
- S/2 values at the two points at a distance l/2 from the center.
- S/5 values at the two points over the edges of the panel and
- Zero subsidence at the two points at a distance equal to 1 from the center.

Thus, a total of seven points will be available and using them approximate subsidence profile can be plotted as shown in Fig. 2.8.

(vi) Slope profile can be plotted by taking the following values.

- Zero slope of the central point,
- G/2 slope at the four points having 0.25 S and 0.75 S values of subsidence,
- G, i.e., maximum slope at the two points having 0.5 S values of subsidence, and
- Zero slope at the two points at a distance l from the center.

Thus, a total of nine points will be available and using than approximate slope profiles can be plotted as shown in Fig. 2.8.

(vii) Strain profile can be plotted by taking the following values.

- Zero strain at the central point,
- Zero strain at the two points at a distance l/2 from the center and also at the two points at a distance l from the center,
- Maximum tensile strain value at the two points of the angles of break, and

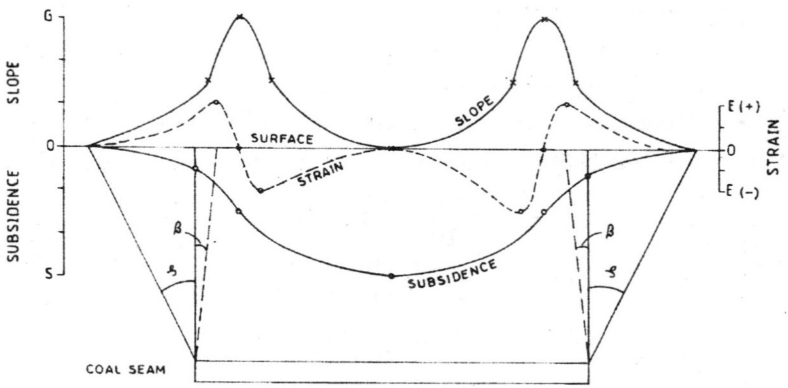

Fig. 2.8. Plotting of subsidence, slope and strain profiles

- Maximum compressive strain value at the two points at a distance equal to that of the maximum tensile strain on the other side of the transition point or the point of half subsidence.

Thus, a total of nine points will be available for plotting the strain profile as shown in Fig. 2.8.

The above predictions of the subsidence movement profiles are generally for critical width-depth ratios, which can also be used for prediction of the subsidence movement profiles for sub-critical and super-critical width depth ratios. In most cases it would suffice to predict only the subsidence profile.

The methodology suggested above predicts symmetrical subsidence movement profiles and in many of the field situations symmetrical profiles are not observed. However, from the point of view of subsidence management planning prediction of symmetrical profiles will suffice. An example of prediction of the subsidence movement profiles is given below.

EXAMPLE 3: Predict subsidence, slope, and strain profiles over a bord and pillar caved panel of 140 m width in a 3.5 m thick seam at a depth of 100 m. The overlying strata consist of interbedded sandstones and shales.

- In this example m′= 3.5 m, h=100 m, w=140 m, M=0.6, w/h=1.4
- Hence, s =25°, ß=7°, e′=0.6 (assuming 80% extraction from panel), a′=0.95, h′=1.0.
- Since the w/h of the panel is 1.4 and M=0.6, the panel is of critical width depth ratio.
- Hence, l=½ [140+ 2.100 tan 20°] = 106.4 m

  Smax=500 (1+ 0.6) (0.6 x 0.95 x 1 x 3.5)= 1596 mm

$$\text{Gmax} = 2\,\frac{1596}{100} = 31.92 \text{ mm/m}$$

$$\text{Emax}_{(-)} = 1.0\,\frac{1596}{100} = 15.96 \text{ mm/m}$$

$$\text{Emax}_{(+)} = 1.05\,\frac{1596}{100} = 16.76 \text{ mm/m}$$

- For subsidence profiles in Fig. 2.9

  x = 0, s = 1596 mm,

  x = ± l/2 = ± 53.2 m, s = 798 mm,

  x = ± w/2 = ± 70 m, s = 319.2 mm, and

$x = \pm l = 106.4$ m, $s = 0$.

Using these values the subsidence profile will be as shown in Fig. 2.9.

- For slope profile in Fig. 2.9

  $x = 0, g = 0,$

  $x = \pm l/2 = \pm 53.2$ m, $g = 31.92$ mm/m,

  $x = \pm 26.6$ & 79.8 m, $g = 15.96$ mm/m,

  $x = \pm l = \pm 106.4$ m, $g = 0$.

The slope profile thus plotted is shown in Fig. 2.9.

- For strain profile in Fig. 2.9.

  $x = 0, e = 0,$

  $x = \pm l/2 = \pm 53.2$ m, $e = 0,$

  $x = \pm 58$ m, $e = E_{(+)} = 16.76$ mm/m

  $x = 48$ m, $e = E_{(-)} = 15.96$ mm/m

  $x = \pm l = 106.4$ m, $e = 0$.

The strain profile is shown in Fig. 2.9.

9. Subsidence, Slope and Strain Contours - Invariably prediction of subsidence contours is done to

- study the changes in surface contours and hence the topography after subsidence, and
- study the changes in surface drainage pattern and possibilities of water logging in the subsided areas.

Prediction of slope and strain contours is seldom done. Hence, the attention in the following discussions will be on prediction of subsidence contours. Various monograms, models, etc. as available in literature on subsidence can be used for prediction of subsidence contours. However, if these are not available the following methodology can be used to plot the subsidence contours manually.

- On the plan of the panel(s), draw lines along and across the length of the panel(s). Care should be taken that there are two lines across the length at a distance of W/2 from the edges as shown in Fig. 2.10.
- On these lines locate the values of subsidence at the points corresponding to those in the subsidence profile as plotted in the example 3 earlier (Fig. 2.9).
- Join the points of equal subsidence to plot the subsidence contours as shown in Fig. 2.10.

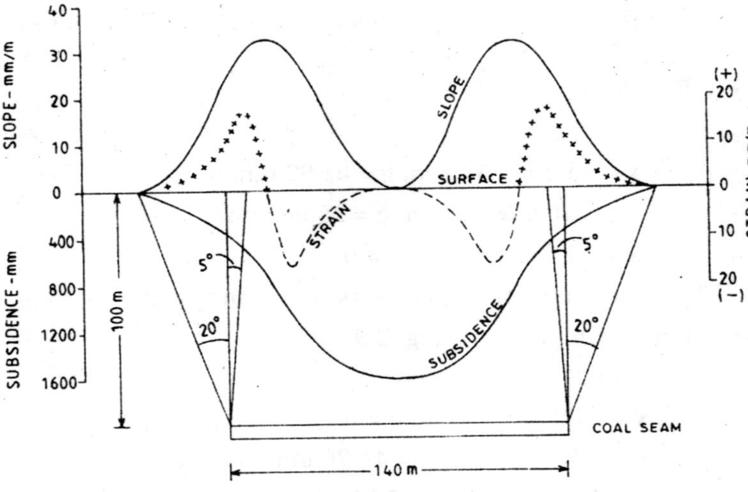

Fig. 2.9. Predicted subsidence, slope and strain profiles (Example 3)

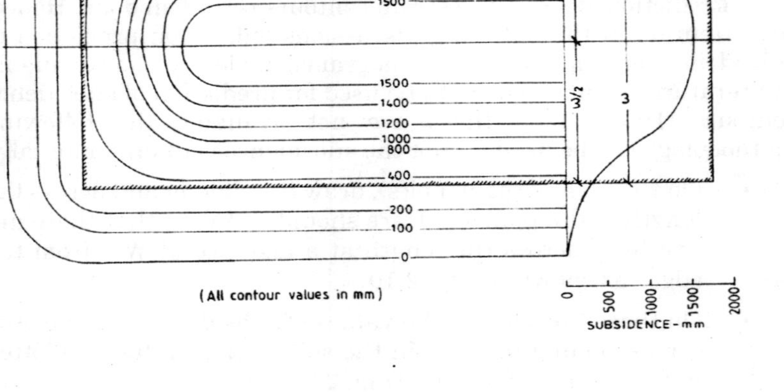

Fig. 2.10. Prediction and plotting of subsidence contours over a panel with the help of predicted subsidence profile

The slope and strain contours, if necessary, can be plotted in the similar manner.

10. Volume of Subsidence Trough : The volume of subsidence trough can be predicted either with the use of the subsidence contours or by using equation (2.10).

$$V = 0.527\ S.A \qquad (2.10)$$

Where, V = volume of subsidence trough in m^3,

S = maximum subsidence in m, and

A = area of underground extraction in m^2.

EXAMPLE 4: Predict the volume of subsidence trough for the extraction of a 1000x150 m panel in a 3.5 m thick seam by longwall caving system at a depth of 100 m, the overlying strata being predominantly sandstones.

In this example m′ = 3.5 m, h = 100 m, L x w = 1000 x 150 m, w/h = 1.5, M = 0.5, e = 100%. Hence, the panel is of critical width-depth ratio. Therefore, the maximum possible subsidence

$$S_{max} = 500\ (1+0.5)\ (0.95x0.95x1.0x3.5) = 2{,}369.1\ mm$$

and $$V = 0.527x2.3691x1000x150 = 1{,}87{,}277.4\ m^3$$

11. Locations of Cracks- The development of tensile strain tends to develop open cracks on the surface and the most likely location for this is the points of maximum tensile strain, which can be marked with the help of the angle of break (ß), as shown in Fig. 2.2. Approximate maximum width of the cracks can be obtained by multiplying the maximum tensile strain value by 15, i.e., a maximum tensile strain of 10 mm/m can be expected to cause about 150 mm wide cracks.

12. Changes in Surface Contours and Drainage Pattern due to Subsidence- Some models available can be used for predicting the changes in surface contours and hence topography due to subsidence. All such models are computer based. In case such models are not available the following methodology can be used.

(i) Divide the layout plan area, i.e., area proposed for extraction and about 80-100 m area around, on a grid pattern with each grid representing 20x20 m to 40x40 m size.

(ii) With the use of the existing surface contours compute the reduced levels of the nodes in the grid.

(iii) Plot the subsidence contours on the grid plan and compute the magnitude of predicted subsidence at the nodes.

(iv) Compute post subsidence reduced levels of the nodes by subtracting the predicted subsidence values from the original reduced level values.

(v) Plot the post subsidence surface contours by using the values of post-subsidence reduced levels of the nodes.

Various graphic softwares available can be used for plotting the post subsidence surface contours as well as changes in the surface topography.

A comparison of the original surface contours/topography and the drainage pattern with the predicted post-subsidence surface contours gives the likely changes in the surface drainage pattern.

The data used in the prediction of the post-subsidence surface contours can also be used in developing pre and post subsidence digital terrain models of the area. An example of this prediction is outlined later.

13. Subsidence Movements at the Locations of Surface Properties - The prediction of the subsidence movements at the locations of the surface properties is done by finding out the magnitude of the movements at the locations from subsidence movement contours. In case of important surface properties it would be appropriate to plot the predicted contours on the plan showing the surface properties.

14. Profiles of Railway Lines, Canals and Roads- Post-subsidence profiles of the railway lines, canals and roads can be predicted by superimposing predicted subsidence along these surface properties on the existing profiles.

The post subsidence profile of the railway lines is studied to assess the long gradients, in distances of about 500 m, which due to the subsidence should not become steeper than the limiting operating gradient.

In case of the canals the changes in the profile due to the subsidence may cause overflow of water from the sides and water logging.

C. PREDICTION OF IMPACTS

Subsidence management planning requires prediction of the impacts of the subsidence movements on surface, sub-surface and underground properties. It is also necessary to predict the problems/dangers of the subsidence movements to the underground workings. These predictions are necessary because the various actions required for subsidence management can be planned so as to take care of the impacts.

1. Problems to Underground Workings- The subsidence movements specially over the caved workings and in some situations over hydraulically stowed workings can cause problems to the underground workings. These are discussed below in brief. It may be noted that so long as the rock mass overlying the extraction area in coal seam is not disturbed due to subsidence movements there are no problems to the underground workings.

(i) **Increased make of water in underground workings**: In both the stowed and caved depillaring processes, there is a possibility of an increase in the make of water in the underground workings due to the subsidence movements.

In stowed cases, although the subsidence movements are less, the maximum subsidence on the surface being about 5% of the extraction thickness, the overlying aquifers and even water table may get disturbed to the extent that water from these sources may find its way to the underground workings. If an aquifer is very close to the coal seam, there may be a substantial increase in the make of water. Also, the stowed goaves can store water up to about 15% of its volume due to air spaces available in the stowed sand and other solids.

In depillaring without stowing the rock mass overlying the goaf breaks and settles in and over the goaf. In this process the aquifers and the water table are also disturbed, as shown in Fig. 1.8, and the water from these sources flows to the underground workings, thereby increasing the make of water in the underground workings. In some situations this may cause severe problems.

It is noted from the above that in both the caved and the stowed mines there are chances of an increase in the make of water and hence during mine planning suitable underground water management measures must be provided.

(ii) **Loss of ventilating air**: As soon as cracks develop on the surface due to the subsidence movements, the underground workings get connected to the surface atmosphere and the air ventilating the depillaring panel finds a way to the surface (Fig. 1.7). This reduces the availability of air in the panel and may cause an alarming increase in the concentration of CH_4 and CO in the return airway of the panel. In order to maintain proper ventilation in the panel, while the leakage to the surface is in progress, it becomes necessary to push more ventilating air into the panel involving extra energy consumption.

(iii) **Development of mine fires**: The ventilating air leaking to the surface through the cracks (Fig. 1.7), specially in bord and pillar depillaring in which a sizable quantity of coal is usually left in the goaves, may cause spontaneous heating of coal left in the goaves, which may ultimately develop into mine fires. Many coal mine fires raging in the Indian coalfields have taken place due to this cause.

(iv) **Increased make of water from surfaces sources**: The development of cracks up to the surface due to the subsidence movements permits rain water and also water from other nearby surface sources under favourable conditions to find way to the underground workings (Fig. 1.3), thus increasing the make of water in the underground workings. In extreme conditions this may cause inundation of the workings.

(v) **Inundation from surface water bodies**: Due to subsidence, the ground may sink below the HFL of surface water bodies and during rainy seasons water from these sources may find its way to the underground workings through the cracks and openings causing inundation.

The methods of mitigation of these problems to the underground workings are discussed later.

2. **Impacts on properties**- The impacts of the subsidence movements on the surface, sub-surface and underground properties can be predicted with the help of the *Subsidence Impact Matrix* (SIM), which is given in *Annexure I*. The predicted values of maximum subsidence, slope and strains are used for the prediction of the impacts. The use of the matrix is discussed below with an example.

EXAMPLE 5: A panel of 150 x 1000 m size in a 3.5 m thick seam is planned for extraction by bord and pillar system with caving with a percentage of extraction of 80 at an average depth of 100 m underneath a railway siding, a few double storied pucca buildings and a road. Predict the problems to the underground workings due to subsidence and impacts on the surface properties.

(i) Assumptions: Rock Mass Factor = 0.6 (for interbedded sandstone & shales).

(ii) Width-depth ratio: 150 ÷ 100 = 1.5, (hence this is a supercritical width panel). Extraction percentage factor = 0.6; Depth factor = 1.0; Goaf treatment factor = 0.95 (for caving).

Hence, Smax = 500(1+0.6)(0.6x 0.95 x l x 3.5) = 1596 mm

$$\text{Gmax} = 2\,\frac{1596}{100} = 31.92 \text{ mm/m}$$

$$\text{Emax}_{(-)} = 1\,\frac{1596}{100} = 15.96 \text{ mm/m}$$

$$\text{Emax}_{(+)} = 1.05\,\frac{1596}{100} = 16.76 \text{ mm/m}$$

(iii) *Problems to underground workings*: 16.76 mm/m of tensile strain will cause about 250 mm wide cracks on the surface and hence all the problems to the underground workings discussed earlier are possible.

(iv) Impacts on surface properties using SIM

a. *A railway siding*: Severe impacts as the predicted maximum compressive and tensile strains are more than 3.0 mm/m and the slope is more than 10 mm/m.

b. *Double storied buildings*: For the predicted values of maximum slope and strains the buildings will be severely damaged and would require major repairs. The buildings will have wide cracks, stepping, crushing, tilting and gaps between the walls and the roof.

c. *A road*: Due to strains more than 10 mm/m the road will need some repairs as these strains will cause about 250 mm wide cracks with some stepping.

D. SUBSIDENCE MANAGEMENT

The subsidence management planning process involves five steps as outlined below.

1. Optimizing subsidence movements to minimize the impacts of subsidence.
2. Impact mitigation.
3. Management of subsided land for various uses.
4. Converting impacts into resources.

5. Monitoring.

1. Optimizing Subsidence Movements: The basic objective of optimization of the subsidence movements is to reduce the impacts of subsidence to the minimum possible extent. This is done by the following.

(i) Reducing the planned extraction thickness.

(ii) Optimizing the size of the panels by reducing their width-depth ratio.

(iii) Planning for alternate stowing and caving depillaring as shown in Fig. 2.11.

(iv) In multiple seam extraction planning for alternative layouts, an example shown in Fig. 2.11.

(v) By adopting stowing methods.

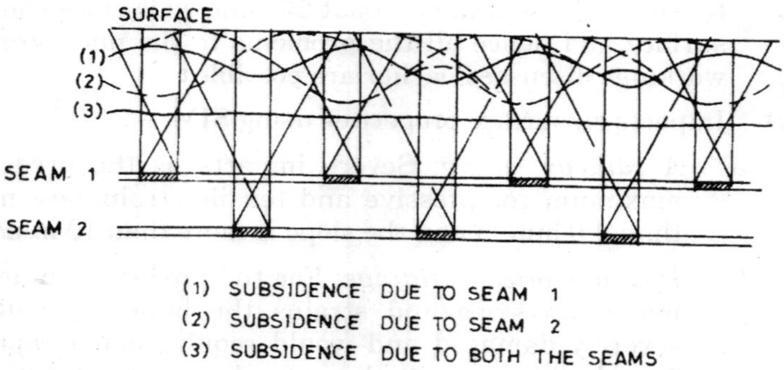

Subsidence trough shape control in two seam mining situation

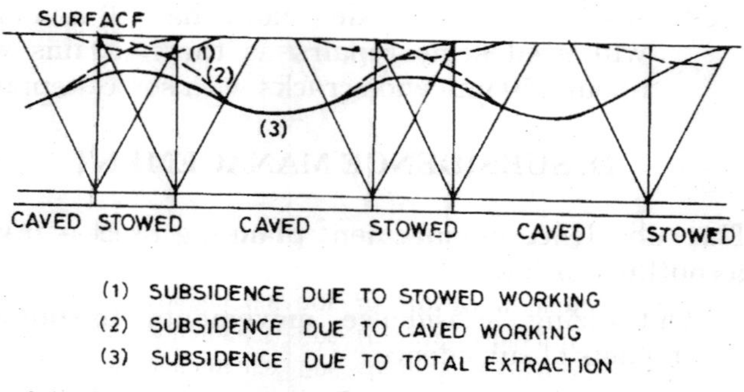

Subsidence trough shape control in single seam mining situation

Fig. 2.11. Alternative extraction layouts for optimising subsidence movements

In the situations where practically no subsidence movements are desirable the following concepts can be used in planning.

(i) Restricting the width of the underground workings so as to bring the width-depth ratio of extraction equal to or less than the non-effective width-depth ratio.

(ii) In bord and pillar mining the panels can be planned for

- splitting as final extraction, and
- widening and heightening of the galleries.

These alternatives are explained hereunder.

(i) Reducing extraction thickness - It is seen from equation (2.3) that under identical conditions of extraction the maximum possible subsidence (Smax) over the underground workings increases with the increase in the extraction thickness (m′)

$$Smax \propto m' \qquad (2.11)$$

Hence, the subsidence movements on the surface can be reduced by reducing the extraction thickness. In practice this would depend on the following.

(i) Seam thickness

(ii) Quality of coal

(iii) Economics of depillaring

(iv) Constraints with respect to equipment, etc.

An example of the application of this method is given in *Annexure II*.

(ii) Optimizing panel size - It is evident from the foregoing discussions that the subsidence movements over the underground workings are maximum when the width-depth ratio of extraction is equal to and more than the critical width-depth ratio. From the critical width-depth ratio to the non-effective width-depth ratio the maximum subsidence movements reduce (Table 2.8). Hence, by reducing the width-depth ratio to a value within the sub-critical range the subsidence movements can be optimized. An example is given in *Annexure III*.

(iii) Planning with stowing- Under identical extraction parameters the maximum possible subsidence (Smax) increases with the increase in the value of the goaf treatment factor (a′), which is 0.07-0.1 for hydraulic stowing and 0.95 for caving. Hence

$$Smax \propto a' \qquad (2.12)$$

Thus, with hydraulic stowing the subsidence is 9.5 to 13.6 times less as compared to that with caving. Therefore, by planning underground extraction with stowing the subsidence movements can be reduced. However, while planning underground extraction with hydraulic stowing the following points need due consideration.

(i) Availability of stowing material

(ii) Additional costs involved in stowing and dealing with stowing water.

(iii) Possible reduction in the rate of production.

(iv) Quality consideration as there may be some mixing of stowing material with the coal/mineral.

Keeping the above points in view hydraulic stowing should be planned only where it is essential. An example of such a planning is given in *Annexure IV*.

(iv) Alternate stowing and caving and barrier-less mining – It is seen from the subsidence impact matrix that the impacts of the subsidence movements on the properties are mainly due to the slope and strains and if these parameters can be optimized by suitable planning methods the impacts on the surface can be minimized. In case of single seam extraction there are two alternatives, namely,

- barrier-less mining, and
- alternate stowing and caving.

In barrier-less mining the extraction planning is done in such a manner that during final extraction no barrier is left between the depillared panels. Two sequences of such a planning are shown in Fig. 2.12. In one sequence the panels are worked in series and the final subsidence profile is as shown in the figure. In the other, the extraction is done in a staggered sequence and in this case also a similar final subsidence profile is obtained. In such a planning it will be necessary to maintain gate roads with goaves on both the sides, which is a rather difficult task, as there is little experience of this nature in India. However, such extractions have been planned and done in other countries.

In some situations it would be appropriate to plan for extraction with caving and hydraulic stowing alternately to optimize subsidence movements on the surface (Fig. 2.11). Extraction with stowing should be planned below the surface properties while with caving it should be planned in the adjacent areas where surface area can be made to subside. Such an extraction was done at Jitpur colliery in Jharia coalfield.

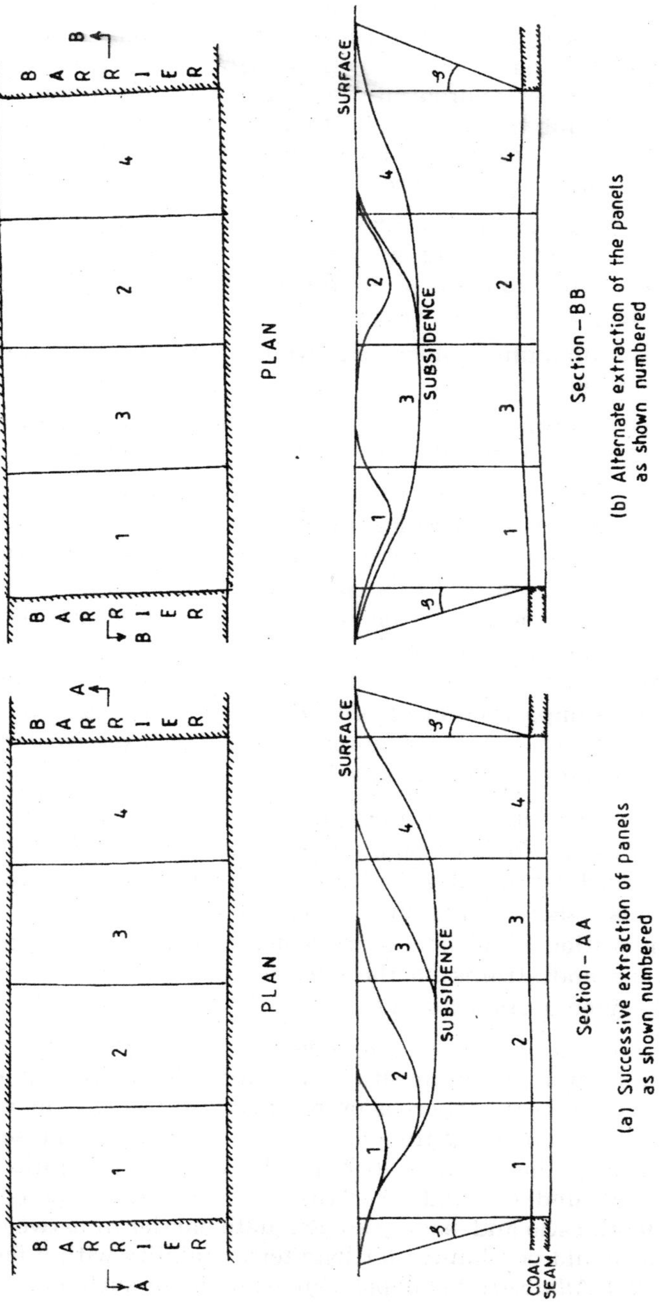

Fig. 2.12. Barrier-less mining layouts

For planning underground mine workings on this pattern, it will be necessary to predict subsidence movements for various combinations of panel widths with stowing and caving and then choosing the most appropriate layout. This method can also be planned in multiple seam situations by predicting subsidence movements. An example illustrating this principle of mine planning is given in *Annexure V*.

(v) Multi-seam extraction with caving- In some situations it becomes necessary to plan multi-seam extraction with caving and also to minimize impacts of subsidence movements. This can be done by adopting the following strategies.

- By planning sub-critical width extraction in superimposition.
- By locating barriers in the seams in such a manner that the slopes and strains on the surface can be controlled.

For planning underground working layouts with sub-critical width it is required to predict subsidence for different sub-critical width-depth ratios of extraction in the seams with the panels in superimposition. After this prediction the impacts of the subsidence movements on the surface are assessed and the most appropriate layout is adopted for planning. In such planning it is necessary that the barriers between the panels in different seams and in superimposition are planned for long-term stability with a factor of safety >2.0. An example of such an exercise for mine planning is outlined in *Annexure VI.*

In many a situations it may not be possible to plan the layout discussed above. In such conditions the subsidence movements can be optimized by suitably locating the barriers in the different seams as shown in Fig. 2.11. These situations cause superimposition of subsidence movements such as to optimize the final slope and strains on the surface. This is explained in the example given in *Annexure VI*.

(vi) Extraction planning with non-effective width-depth ratio – Many underground mining situations warrant extraction planning in such a manner that there is practically no subsidence movement on the surface and/or at the plane of reference above the seam(s). This can be achieved by restricting the width-depth ratio of the panels in underground workings up to the non-effective width-depth ratio and leaving an adequate barrier in-between. The barriers should be planned for long-term stability with a factor of safety >2.0. *Annexure VII* depicts an example for such a planning.

The planning process involves assessment of the strength and factor of safety of the reduced pillars using various equations available. An example is discussed in *Annexure VIII*.

(vii) Partial extraction of bord and pillar developed areas - Partial extraction of the pillars is adopted in situations where practically no subsidence movement is desired on the surface or the level above the seam. In these methods the pillars after development are split into two, four, six and even up to nine parts by splitting as shown in Fig. 2.13, and in some situations the overall size of the pillars is reduced by widening and heightening the galleries as shown in Fig. 2.14. The design criteria envisages that the stooks/pillars after reduction have adequate factor of safety, which is taken as 1.0 and more for extraction planning with hydraulic stowing and more than 2.0 for caving conditions.

Although the partial extraction methods can be planned without stowing as such extractions have been done in past, presently such extractions are not being permitted without stowing.

2. Impact Mitigation

Once the subsidence movements have been optimized and impacts have been predicted the process of impact mitigation planning is taken up. Invariably, the impact mitigation planning takes into account the intended use of surface land during and after mining. It involves the following steps.

- Outlining the locations where cracks are likely to develop on the surface and the likely changes in surface contours.
- Study of the impacts on surface properties.
- Mitigation planning for minimizing dangers to underground workings.
- Mitigation of impacts on the surface properties.

(i) Locations of the cracks on the surface can be predicted with fairly good accuracy by adopting the methodology outlined earlier in this chapter and these locations can be marked on the plans. The likely width of the cracks can also be predicted and indicated on the plans.

(ii) Changes in surface contours due to subsidence can be predicted by the methodology outlined in this chapter and the changed contours can be plotted on the plans for studying the

- changes in surface drainage pattern, and

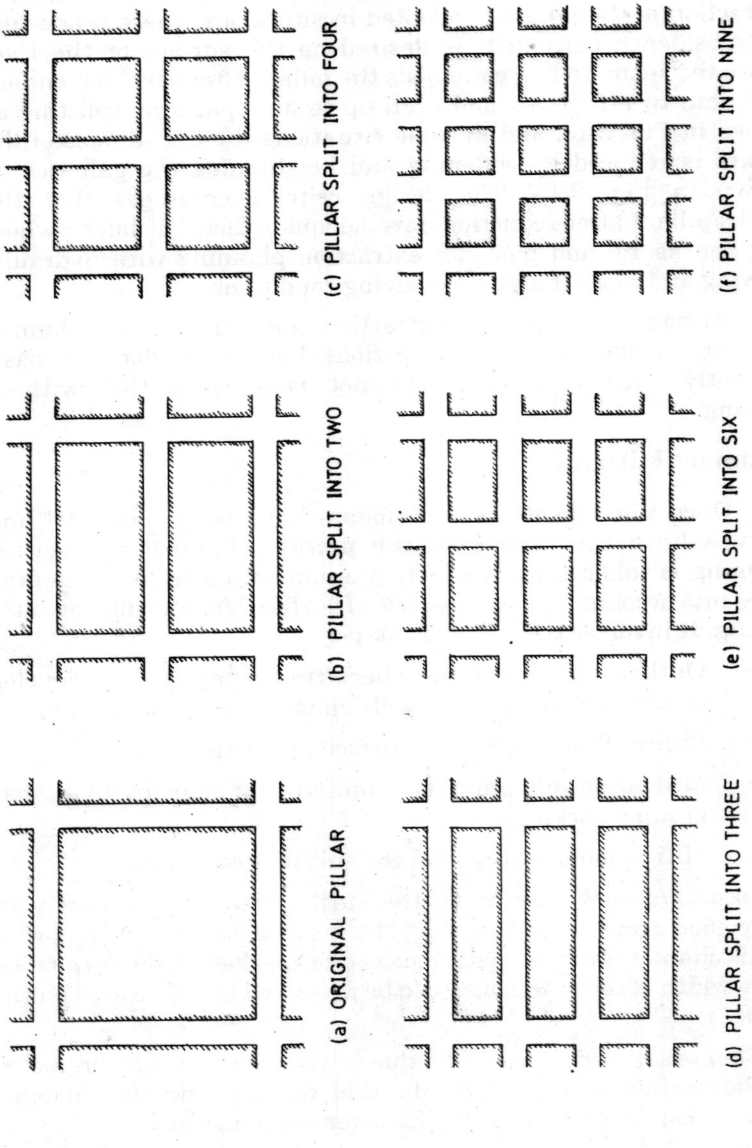

Fig. 2.13. Splitting of pillars

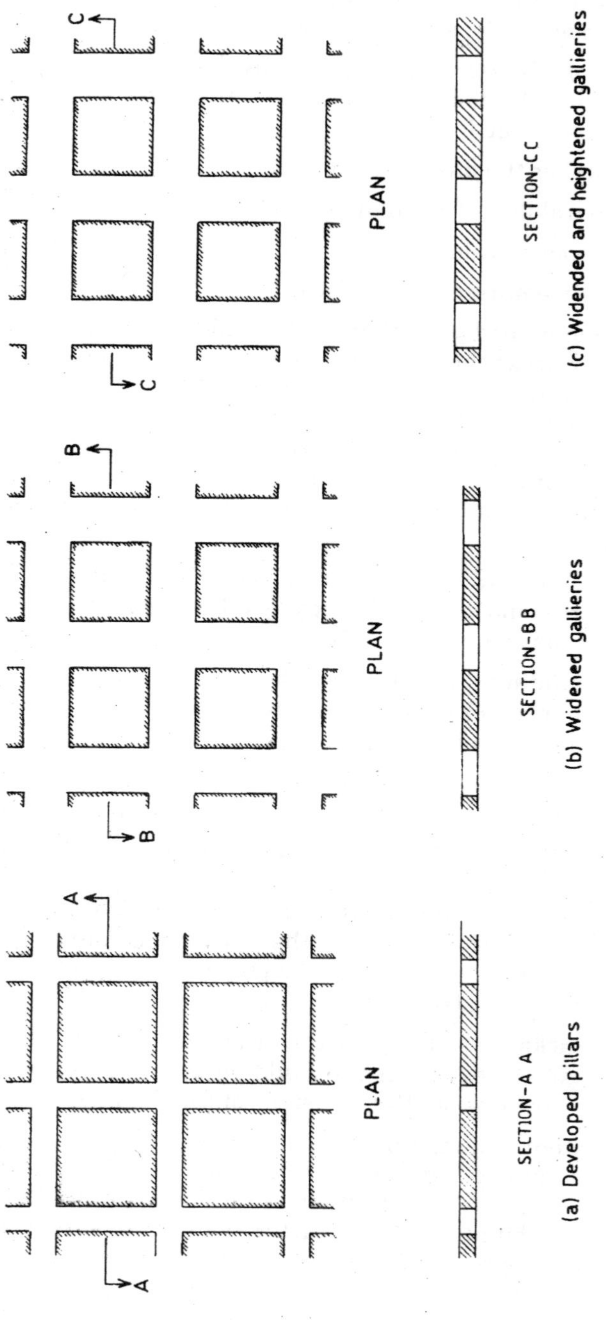

Fig. 2.14. Widening and heightening of gallieries

- possibilities of water logging on the surface due to the changes in the surface contours.

(iii) Study of the impacts of the subsidence movements – The impacts of the subsidence movements on the surface properties are studied to assess the following.

- Whether the damages are temporary or permanent.
- Reparability of the damages.
- Economic evaluation.
- Possible mitigative measures.

It may be mentioned here that in the buildings and structures in the subsiding areas the damages are typical due to excessive slope, compressive strains and tensile strains. Also, all the properties (buildings and structures) should theoretically exhibit similar nature of damages in the tensile strain zone and also those in the compressive strain zone. Isolated damages in the buildings and structures are usually due to causes not related to subsidence movements.

The damages due to slope or tilt comprise of tilting of building and structures and the walls may exhibit distinct displacement with respect to the roof (Fig. 2.15).

The buildings in the tensile strain zone may exhibit development of cracks which may widen with height above the plinth area as shown in Fig. 2.16.

In the compression zone there may be lifting and overriding in the floor, tight cracks in the walls with crushing of wall material and these cracks may or may not extend up to the roof (Fig. 2.17).

Invariably subsidence movements cause irreversible but repairable damages in the buildings and structures. However, depending upon the magnitude of the slope and strain the damages may be rated as severe to very severe in which case it may be required to raze the structure.

The cracks developing in the travelling tensile strain zone sometimes get closed when the buildings/structures come in the compressive strain zone. But the signs of the damages remain.

The extent and magnitude of the damages also depend on the nature, size and manner of construction of the buildings and structures. In many a situations in the Indian coalfields the damages to the buildings and structures due to reasons other than subsidence were projected as those having taken place due to the subsidence movements.

Fig. 2.15. Damage to a building due to slope.

Fig. 2.16. Cracks in a building due to tensile strain

Fig. 2.17. Damage to a building due to compressive strain

Economic evaluation of the damages involves assessment of the

- expenditure involved in the repair work,
- if buildings/structures are to be demolished, then cost of demolishing,
- if people are to be rehabilitated, then the cost of rehabilitation,
- in case of the loss of crops, the crop compensation, etc. to be paid,
- expenditure involved in augmenting water supply, etc.

Sometimes it may become essential to construct new residential buildings and in such situations the costs on the new constructions are also required to be assessed. The costs evaluated are to be added to the project cost and should be reflected in the cost benefit analysis.

(iv) Minimizing dangers to the underground workings – A study of the dangers to the underground workings due to subsidence movements, described earlier, indicates that the dangers are mainly due to

- development of the cracks up to the surface,
- lowering of ground level below the HFL, and

- disturbances in the overlying aquifers and water table.

The development of the cracks up to the surface causes the loss of ventilating air from the depillaring districts, chances of development of mine fires, and increased make of water and may also lead to inundation from surface sources. These dangers can be minimized simply by filling up the cracks on the surface so that

- the channels for leakage of air from the underground workings to the surface are closed, and
- also the channels for water to go to the underground workings from the surface sources are closed.

Hence, the filling of the cracks on the surface should be started as soon as they start developing. The cracks should be filled with a mix of stone chips and soil tightly and after filling a small heap of soil be made over the filled cracks as shown in Fig. 2.18. In the situations having more than 300 mm wide cracks with stepping it may be appropriate to plan for dozing the surface soil before effectively filling the cracks. While planning the mine, provisions should be made for these activities and necessary costs on manpower, equipment, etc. be accounted for. In the mines to be planned with caving methods filling of cracks and dozing are invariably required continually once depillaring is started. The costs on these mitigative measures should invariably be compensated by the significant improvement in underground ventilation and reduction in pumping. In fact, filling of the cracks and necessary dozing on the surface should be treated as a part of the mining activities.

For minimizing the chances of inundation of the underground workings from surface sources due to lowering of the ground below the HFL of the adjoining water bodies the following mitigation steps are to be planned.

- Study of the post subsidence surface contours (at 1.0 m intervals) to demarcate the area coming below HFL.
- Study of the possible connections of the area below HFL to the mine openings and thus assessment of possible water channels reaching the openings.
- Study of the surface contours for the assessment of the locations where embankments can be made to stop flow of water to the mine openings.
- Selection of the most suitable locations of the embankments.
- Designing the embankments with an appropriate factor of safety.

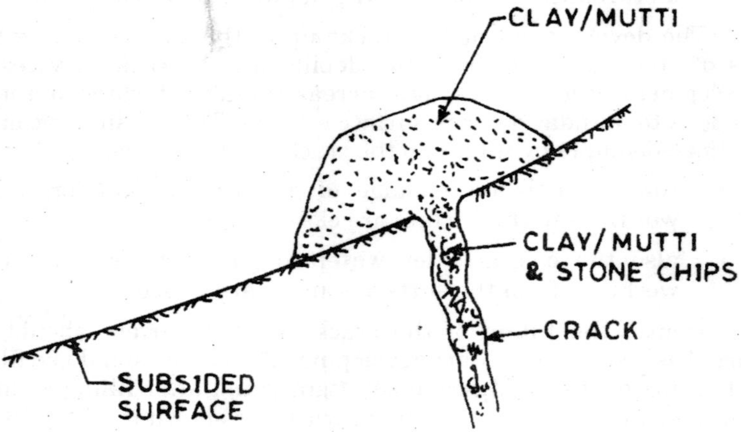

Fig. 2.18. Filling of a crack

This exercise requires up-to-date accurate surface plan with surface contours at the interval of 1.0 m. In case of any doubt a fresh survey may be planned for this purpose. Sometimes, for erecting the surface, infrastructure changes in the surface topography may be required. These changes should also be taken into consideration. Costs involved in making the bund(s) or embankment(s) should be incorporated in the project costs. Design of embankments should take into consideration all the civil engineering, soil mechanics and hydrological parameters.

Underground mining in sedimentary deposits disturbs the overlying aquifers and water table and the water from these sources may find a way to the underground workings and this may happen with or without the development of the cracks up to the surface. The rate of water coming to the underground workings from these sources depends on the following.

- Characteristics, e.g., thickness, porosity, permeability, piezometric head, etc., of the aquifers and water table.
- Extent of disturbances in the aquifers and water table.
- Connectivity of these underground water sources with the surface water sources.

With the present status of technology it is not possible to predict the make of water in the underground workings accurately. However, approximate assessment can be made by hydrological studies.

In the situations where higher make of water is anticipated in the underground workings during depillaring the following measures can be planned.

- The development should be planned in such a manner that there is enough room available on the dip side of the depillaring districts to store the incoming water for about 24 to 32 hours.
- The depillaring districts should be suitably connected to the dip side workings so that the water coming in these districts finds a way to the dip side workings.
- There should be adequate pumping capacity in the mine to pump out water. Sometimes it may be appropriate to install pumps in the dip side workings to take out water through bore-holes made from the surface.

The costs of making the above arrangements and operating costs of the pumps should be incorporated in the project cost.

3. Management of Subsided Land for Various Uses

In the countries like India land is an important resource and hence every inch of land available should be properly used. This is also true for the subsided land. As is well known most of sedimentary mineral deposits in India are in forest-cum-agricultural land, where land is important. Hence, it would be appropriate to plan for minimum possible loss of land use in subsided areas. The subsided land can be managed for various uses in the manner outlined hereunder.

(i) Developing surface water bodies — Both temporary as well as permanent surface water bodies can be developed in the subsided areas and planning for the development of these water bodies can be done in the following manner.

Permanent surface water bodies are planned to be developed towards the end of the mining activities such that the area is not subjected to practically any further subsidence movement. Temporary water bodies can be planned during the mining activities with provisions for draining them as and when required during mining.

For both the developments it is required to predict post mining surface contours at least at 1.0 m interval. This can be done in the manner outlined earlier. Once these contours are available they are studied to identify the depressions developed due to the subsidence movements and their suitability for the development of surface water bodies. Thereafter, the design of the water bodies is done which in most cases may involve design and planning of suitable embankments at appropriate locations. The design of the embankments should be made on the considerations outlined earlier.

In the cases of the development of the temporary surface water bodies it would be appropriate to incorporate the drainage methodology in the design itself.

Development of surface bodies is required for water management and augmenting water resources in mining areas. Since the availability of water from the underground sources in the nearby surrounding areas tends to reduce the water available from the surface water bodies can be used to compensate the losses. The following points need due consideration for the development of such water bodies.

1. Accurate surface plan with surface contours at least at 1.0 m interval.
2. Accurate correlation of the underground workings with the surface.
3. On the basis of 1 and 2 prediction of post subsidence surface contours at 1.0 m interval.
4. Prediction of the locations where cracks are likely to develop on the surface and the approximate size of the cracks.
5. Filling of the cracks adequately.
6. Design of bunds/embankments for long-term stability and leak proofness.
7. Disinfecting the area before formation of the water body.
8. No further subsidence possibility in case of the permanent water bodies.
9. Arrangements for drainage in case of the temporary water bodies.
10. Removal of all vegetation from the designated area.
11. Any other precaution that may be required based on local situation and intended use.

(ii) Developing forestry, plantation and grassland – The subsided areas can be used for developing forestry, plantation and grassland at every stage of mining underground. The steps involved in planning are as outlined below.

1. Predict subsidence movements stage-wise.

2. Predict development of the cracks and develop post mining surface contours.
3. Identify areas likely to have possible accumulation of water in rainy season.
4. Plan for adequate filling of the cracks.
5. Invariably modifications in the post subsidence surface topography are not required for developing forestry, plantation and grassland. The water bodies developing due to subsidence can be used in dry seasons.
6. Forestry, plantation and grassland can be developed by following normal practices.

(iii) Development of subsided land for agricultural purposes can be done in the manner outlined later. Emphasis should be made that the agricultural land use should continue while subsidence is taking place except in the cases where the width of the cracks due to subsidence is >500 mm.

1. Predict subsidence movement stage-wise, development of cracks, and post subsidence surface contours.
2. Identify areas likely to have water logging in rainy seasons.
3. Plan for suitable filling of the cracks.
4. Plan for suitable drainage of the area so as to minimize water logging. Some modifications in the surface topography may be required for this purpose.
5. Once crack filling and drainage has been planned agricultural use of land can be planned in such a manner that the land remains unused for minimum possible time.

In the above planning, filling of the cracks improves water retaining capacity of the top soil and drainage arrangements make the land available for use.

(iv) Development of subsided land for construction purposes becomes necessary in large coalfields where non-coal bearing areas are not available for this purpose. The constructions could be temporary as well as permanent. The temporary constructions required for about 5 to 15 years time can be planned in the manner outlined hereunder.

1. Predict stage-wise subsidence movements, location of cracks and post mining surface contours.
2. Identify areas likely to become waterlogged during rains.

3. Identify areas not likely to have any further subsidence for the period for which the construction is to be made. If necessary, some modifications in the underground extraction plan can be done for this purpose.
4. Once the areas are identified study their drainage pattern and plan for adequate drainage to avoid water logging.
5. Plan for adequate filling of the cracks.
6. Assess the load bearing capacity of the soils at foundation level and design the constructions accordingly.
7. Under no circumstances the individual building should be of more than two story (about 7 m above ground) and 20 m long.
8. Necessary civil engineering designs be taken into consideration for such temporary/semi-permanent constructions.

Planning for permanent buildings and structures should be done over the areas not likely to have any further subsidence once construction work has started. The following steps are involved in such a planning.

1. Predict subsidence movements, locations of cracks and post mining subsidence contours.
2. Ensure that there is no further chance of subsidence in the area.
3. Identify areas likely to become waterlogged during rainy seasons.
4. Plan for adequate filling of the cracks.
5. Plan for adequate drainage of the area.
6. Assess the post subsidence load bearing capacity of soils at the foundation level and design constructions accordingly.
7. Incorporate civil engineering design features in the constructions and with suitable precautions even multi-storied constructions can be planned.

(v) Construction of roads over subsided areas can also be planned and for this purpose the following steps are involved in planning.

1. Predict subsidence movements, location of cracks and post-mining subsidence contours.

2. Identify the path for the road over the subsided area.
3. Identify stretches in the path likely to become waterlogged and plan for adequate drainage of these areas.
4. Plan for adequate filling of the cracks.
5. Plan for laying of the road as per normal road construction practices.

Similarly, planning can be done for laying of railway lines, canals and pipelines through the subsided areas and in these cases the additional major requirements are no further subsidence potential and normal restrictions in their construction.

4. Converting Impacts into Resources

Under favourable conditions the adverse impacts of the subsidence movements can be converted into beneficial ones. The best example of this nature in the world is increasing the draft of water in Duisburg inland harbour in Germany by planned subsidence due to underground mining of coal seams. Another example being slight flattening of the long gradient of a main railway line due to subsidence over the stowed workings of Sudamdih Shaft and Incline mines in Jharia coalfield. Hence, depending upon the local conditions on the surface and the magnitude and impacts of the subsidence movements the adverse impacts can be converted into advantageous ones.

The author opines that, in favourable conditions, the following resources can be effectively developed in subsided areas.

(i) Surface water bodies — The manners in which temporary and permanent surface water bodies can be developed in the subsided areas have been discussed in detail earlier. Hence, this is not being outlined here.

(ii) Underground water bodies — In underground mining in sedimentary deposits the total volume of extraction underground does not get reflected in the volume of the subsidence troughs. In fact the volume of the subsidence troughs seldom exceeds 50% of that extracted underground in caved workings. Even in stowed workings a certain percentage of void (about 15%) remains in underground stowed areas. These voids can be filled with water while decommissioning the mine and arrangements can be made for charging the workings with water and also for the withdrawal of the water from the underground water bodies thus formed. This is discussed with a hypothetical example below.

(i)	Area of mine	=	10 sq km
(ii)	Number of seams	=	2
(iii)	Extraction thickness	=	3 m (for each seam)
(iv)	Depth	=	200 m
(v)	Overall percentage of extraction		
	• Developed area	=	30%
	• Depillared area	=	70%
	• Extraction in developed area	=	25%
	• Extraction in depillared area	=	80%
(vi)	Total volume of voids in developed area	=	4.5×10^6 m^3
(vii)	Total volume of voids in depillared area	=	33.6×10^6 m^3
(viii)	Volume of voids available in underground workings in	=	
	a. development & depillaring with caving	=	17.3×10^6 m^3
	b. development & depillaring with stowing	=	6.55×10^6 m^3
(ix)	Quantity of water that can be stored in underground in condition with		
	a. caving	=	17.3 billion litre
	b. stowing	=	6.55 billion litre
(x)	At a rate of supply of 200 l/head/day these quantities will be sufficient for a population of about	=	2.37 lakh for caved condition and 0.9 lakh for stowed condition for round the year supply.

Formation of underground water bodies have the following advantages and disadvantages compared to the surface water bodies.

Advantages

(i) Practically no loss of water due to evaporation.

(ii) Water is not likely to get polluted due to human activities on the surface.

(iii) Round the year availability of water can be ensured.

(iv) Aquifers and water table damaged due to mining may get rejuvenated and availability of water in the surrounding areas may improve from these sources.

Disadvantages

(i) Water may acquire dissolved and suspended solids from the rocks.

(ii) There may be some minor subsidences on the surface due to readjustments underground.

For the formation of the underground water bodies the following steps are to be incorporated in the decommissioning plan part of the project report.

1. Predict subsidence movements and the locations and size of the surface cracks likely to develop due to subsidence.
2. Adequate filling of all the cracks.
3. Adequate closing of all the mine openings, i.e., shafts and inclines.
4. Putting bore holes from the surface to leftover openings and placing pipes, with perforations at both ends in the bore holes in the manner shown in Fig. 2.19.
5. Installing bore hole pumps for the withdrawal of water in the manner shown in Fig. 2.19.

The locations of the bore holes for feeding the water and also for its withdrawal are to be chosen on the basis of the layout of the underground mine workings and surface topography and layout.

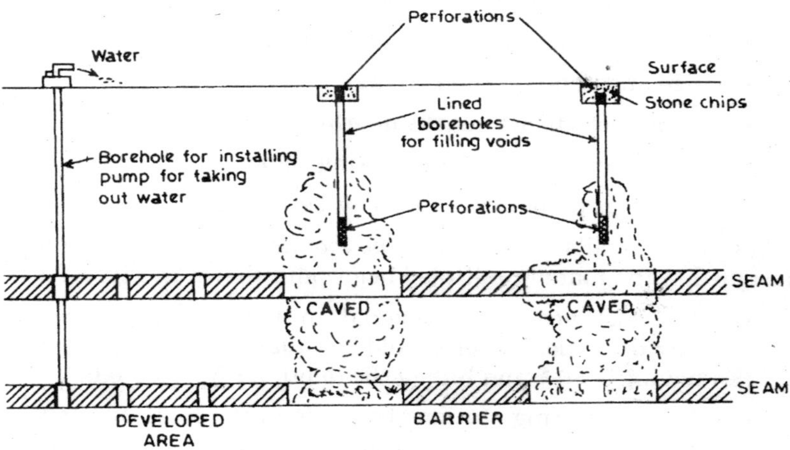

Fig. 2.19. Developing an underground water body

E. MONITORING OF SUBSIDENCE MOVEMENTS

The main *objectives* of monitoring of the subsidence movements are

- to check the accuracy of the predictions and if necessary to do modifications,
- to assess actual impacts of the subsidence movements, and
- to implement the mitigative measures effectively.

The *monitoring aims at* getting the following data in respect of the subsidence and impacts.

- Maximum subsidence, slope and strains
- Angle of draw and angle of break
- Subsidence, slope and strain profiles
- Subsidence contours
- Volume of subsidence trough
- Location of surface cracks
- Changes in surface contours due to subsidence
- Impacts of the subsidence movements

The *data required for planning the monitoring* of the subsidence movements are

- an accurate surface plan showing accurate locations of the proposed underground workings,
- surface contours at interval of 1.0 m,
- size and depth of the panel(s),
- permanent bench marks nearby,
- adjoining underground workings, and
- workings (depillared) above or below.

Using these details *the layout of the monitoring stations* should be designed in the manner outlined below and shown in Fig. 2.20.

- The distance to be covered in monitoring beyond the extraction perimeter should be equal to the depth for strata having M>0.7 and half of the depth for M = 0.5-0.6.
- The monitoring stations be made in the lines as shown in Fig. 2.20 to cover strike, dip-rise and diagonal directions.
- The interval between the monitoring stations should be 1/20 of the depth between 5 and 15 m.

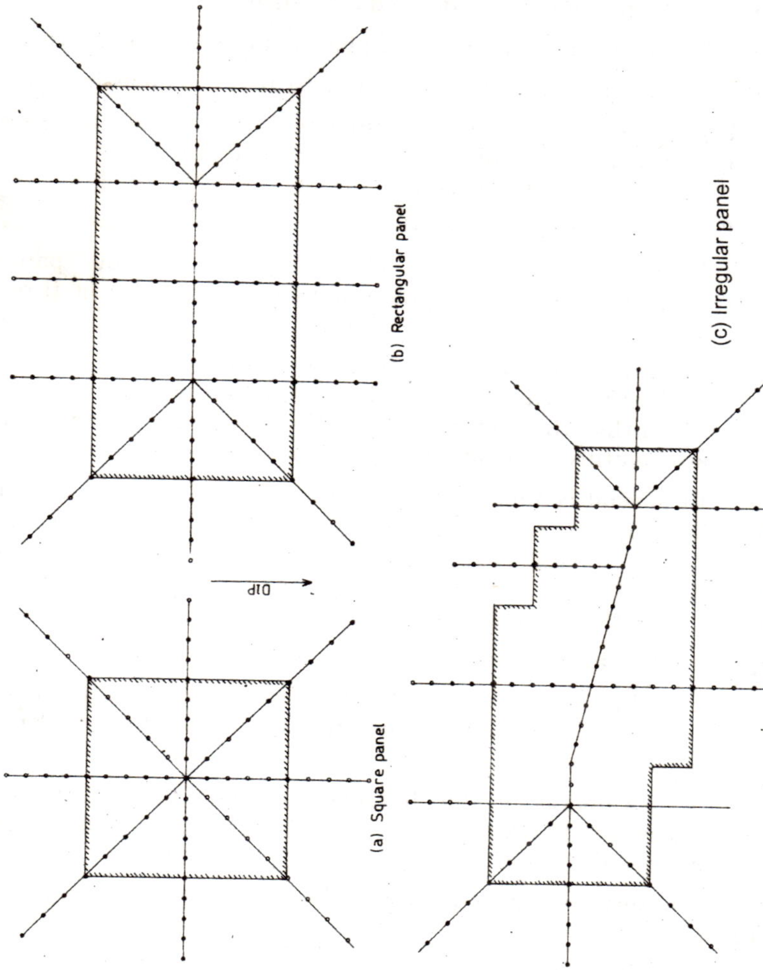

Fig. 2.20. Layouts of monitoring stations

- The layout is to be connected to bench mark(s) with stations at intervals of 15-20 m.

The monitoring stations should be made to *conform to the following*.

- These are not easily disturbed by men and animals.
- These provide a well defined reference point.
- These afford exact repeated placement of leveling staves.
- These afford accurate intermediate distance measurement.

The *best design* should have a steel ball with a punch mark on the top (Fig. 2.21). It would be better to ensure that the distances between the stations are measured directly.

The *monitoring frequency* should be as outlined below.

- Before commencement of extraction in the panel underground at least two readings at intervals of three months should be taken.
- In the normal surface conditions monitoring should be done quarterly during the period of extraction and up to about 6-months thereafter. Then the frequency should be six-monthly for another two to three years.
- In special conditions, e.g., working below railway line or built-up areas, monitoring should be monthly to alternate month during the period up to about 6-months after completion of extraction in the panel followed by quarterly for another two to three years.

The following *operations* are done in monitoring.

- Levelling with an accuracy of 0.1 mm using parallel plate micrometer fitted levels to obtain the reduced levels (RLs) of the points in the layout.
- Measurement of the distances between the monitoring stations with an accuracy of 1.0 mm using steel bands with standard tension.

The observation are recorded in the *field books* in the manner outlined below.

- Table 2.13 gives a sample page of field book recording for leveling.
- Table 2.14 gives a sample page of field book recording for distance measurement.

The *computation of the subsidence movements* from the

Fig. 2.21. Some views of subsidence monitoring stations

observations of leveling and distance measurement is done in the manner outlined in the following tables.

- Table 2.15 gives a sample of subsidence computation.
- Table 2.16 gives a sample of strain and displacement computation.
- Table 2.17 gives a sample of slope and curvature calculation.

Tables 2.13 to 2.17 are self explanatory and can be easily adopted.

Table 2.13 : Sample field record for leveling

Station number	Sight	Left hand reading - A	Right hand reading - B	Difference (A-B)	Rise	Fall	Reduced level	Remarks
		m	m	m	m	m	m	
100	BS	3.1000	3.4020	0.3020			100.0000	
101	IS	3.2300	3.5320	0.3020	0.1300		100.1300	
102	IS	3.4600	3.7620	0.3020	0.2300		100.3600	
103	IS	3.0100	3.3120	0.3020		0.4500	99.9100	
104	IS	3.0050	3.3070	0.3020		0.0050	99.9050	
105	IS	2.9000	3.2020	0.3020		0.1050	99.8000	
106	IS	2.8000	3.1020	0.3020		0.1000	99.7000	
107	IS	3.0800	3.3820	0.3020	0.2800		99.9800	
101	FS	3.1000	3.4020	0.3020	0.0200		100.0000	

BS : Back sight; IS : Intermediate sight; FS : Foresight

Table 2.14 : Sample field recording for distance measurement

Bay	Left hand reading	Right hand reading	Height of stand on LHS	Height of stand on RHS	RL of LHS station	RL of RHS station	Inclined distance	Difference of level	Horizontal distance
	m	m	m	m	m	m	m	m	m
A1-A2	0.000	10.255	0.000	0.000	100.250	100.450	10.255	0.250	10.252
	0.000	10.255	0.000	0.000	100.250	100.450	10.255	0.250	10.252
A2-A3	0.255	10.460	0.350	0.400	100.450	100.500	10.205	0.195	10.203
	0.460	10.665	0.350	0.400	100.450	100.500	10.205	0.195	10.203
A3-A4	0.545	10.605	0.400	0.650	100.500	100.650	10.060	0.400	10.052
	0.680	10.740	0.400	0.650	100.500	100.650	10.060	0.400	10.052
A4-A5	0.600	11.450	0.650	1.250	100.650	99.600	10.850	0.450	10.841
	0.750	11.600	0.650	1.250	100.650	99.600	10.850	0.450	10.841

Table 2.15 : Subsidence computation

Station number	Original RL on 1.1.2000	RL on 1.4.2000	Subsidence	RL on 1.7.2000	Subsidence	RL on 1.10.2000	Subsidence
	m	m	m	m	m	m	m
A1	100.2500	100.2500	0.0000	100.2500	0.0000	100.2500	0.0000
A2	100.4500	100.4450	0.0050	100.4200	0.0300	100.4100	0.0400
A3	100.5000	100.4900	0.0100	100.4500	0.0500	100.4000	0.1000
A4	100.6500	100.6100	0.0400	100.5500	0.1000	100.4000	0.2500
A5	99.6000	99.4000	0.2000	99.2000	0.4000	99.0000	0.6000
A6	100.0200	99.7000	0.3200	99.5000	0.5200	99.2000	0.8200
A7	100.3500	99.5000	0.8500	99.2000	1.1500	99.1000	1.2500
A8	99.8800	99.1000	0.7800	98.7000	1.1800	98.5500	1.2500

Table 2.16 : Strain and displacement computation

Bay	Distance on 1.1. 2000	Distance on 1.4. 2000	Displacement	Strain	Distance on 1.7. 2000	Displacement	Strain	Distance on 1.10. 2000	Displacement	Strain
	m	m	mm	mm/m	m	mm	mm/m	m	mm	mm/m
A1-A2	10.252	10.252	0	0.0	10.255	+3	+0.3	10.260	+8	+0.8
A2-A3	10.203	10.208	+5	+0.5	10.215	+12	+1.2	10.220	+17	+1.7
A3-A4	10.052	10.060	+8	+0.8	10.070	+18	+1.8	10.080	+28	+2.8
A4-A5	10.841	10.840	-1	-0.1	10.845	+4	+0.4	10.845	+4	+0.4
A5-A6	10.500	10.490	-10	-1.0	10.485	-15	-1.4	10.480	-20	-1.9
A6-A7	10.650	10.630	-20	-1.9	10.625	-25	-2.3	10.620	-30	-2.8
A7-A8	10.550	10.510	-40	-3.8	10.500	-50	-4.7	10.480	-70	-6.6
A8-A9	10.750	10.710	-50	-4.7	10.690	-60	-5.6	10.650	-100	-9.3

+Tensile strain, Compressive strain

Table 2.17 : Slope and curvature computation

Station number	Original RL on 1.1.200	RL on 1.10. 2000	Original distance	Subsidence	Difference of subsidence	Slope	Difference in slope	Curvature	Radius of curvature
	m	m	m	m	m	mm/m		m-1	m
A1	100.250	100.250		0.000					
			10.252		0.040	3.9			
A2	100.450	100.410		0.040			0.0020	0.00020	5000
			10.203		0.060	5.9			
A3	100.500	100.400		0.100			0.0090	0.00090	1111
			10.052		0.150	14.9			
A4	100.650	100.400		0.250			0.0185	0.00185	540.5
			10.841		0.350	32.4			
A5	99.6000	99.0000		0.600			0.0114	0.00114	877.2
			10.500		0.220	21.0			
A6	100.020	99.2000		0.820					

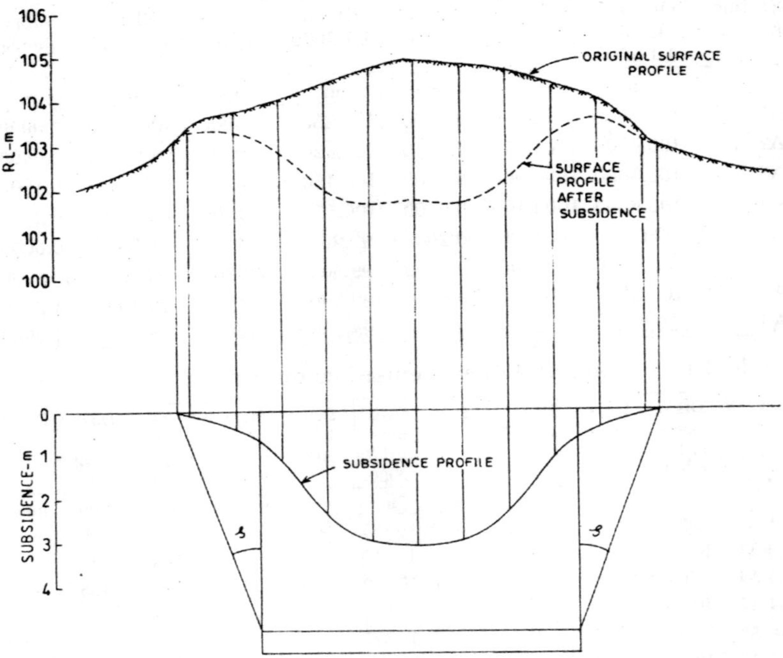

Fig. 2.22. Changes in surface profile due to subsidence

The *subsidence profile and the changes in the surface profile* can be plotted in the manner illustrated in Fig. 2.22. The subsidence profile is used to compute the angle of draw and the strain profile is used to compute the angle of break.

The *subsidence contours and post subsidence surface contours* are plotted with the use of the subsidence values at the monitoring stations and the post subsidence RLs of the monitoring stations respectively.

The *volume of the subsidence trough* is computed in the manner outlined below.

- Areas of all the subsidence contours from zero to maximum subsidence are computed.
- Using the areas of the subsidence contours and the difference in subsidence values between successive contours the volume is computed. An example is given in *Annexure IX*.

If suitable computer softwares are available the above computations and plotting can be done with their use. Manual computation and plotting is also not likely to take much time.

1. Case Studies

The author has conducted more than 100 subsidence case studies for the following purposes.

1. For underground extraction of coal seams below surface, sub-surface and underground properties.
2. For developing the Environmental Management Plan(s) of mining projects.
3. For planning underground extraction below forest areas.

Invariably in such case studies prediction of maximum subsidence movements and their impacts is required. Rarely there is a necessity of predicting post subsidence surface contours, etc. The subsidence prediction requirements are described in Table 2.18. Some important case studies are outlined below.

Table 2.18 : Subsidence prediction requirements

1	For planning extraction below surface properties including water bodies	Maximum subsidence, slope and strains.
2	For planning extraction below railway lines, roads, pipe lines, etc.	Maximum subsidence, slope, strain and changes in the profiles of the railway line, etc. due to subsidence.
3	For mining below forest areas	Maximum subsidence, slope and strains. Computation of tilt of trees and width of cracks.
4	For changes in surface drainage pattern and management of surface for developing surface water bodies, etc.	Post mining surface contours of 1.0m intervals and location and width of the cracks.

I. Sudamdih Shaft and Incline Mines of Bharat Coking Coal Limited in Jharia Coalfield

In the beginning of the 1970s it was proposed to extract XI/XII Seam (7m thick) by longwall system with hydraulic sand stowing below a main railway line at Sudamdih Incline Mine at depths ranging between 35 m and about 100 m. After anticipating maximum subsidence, slope and strains it was recommended that one slice could be extracted between the depths of 35 m and 60 m and two slices at depths >60 m. The following safe limits were taken into consideration for the railway line of jointed construction.

Maximum strain = 3 mm/m

Limiting operating gradient for the railway line = 1 in 100.

Accordingly extraction was started in the seam. As mining progressed towards the dip side it was noticed that the existing gradient of the railway line was slightly flatter than 1 in 100, the limiting gradient, in some part. Hence, the changes in this gradient were predicted. In 1975 it was found that there are old water logged workings in three overlying seams, namely, XIV, XIVA and XV, and the parting between the XI/XII and XIV seam was about 100 m. Hence, the width of extraction in XI/XII seam panels was restricted to 30 m, i.e., 0.3 times the parting.

Later it was proposed to extract VIIIA, IX/X and XI/XII seams also with stowing at the Shaft Mine. For studying the feasibility of extraction the subsidence movements likely to occur due to extraction of the individual seams and also collectively were anticipated. The changes in the profile of the railway line were also anticipated. Based on these calculations recommendations were made for the extraction of the coal seams.

The extraction below and in the vicinity of the railway line continued till mid 1990s and during the entire period subsidence movements were regularly monitored along the railway line along with the profile of the railway line. These observations were periodically reviewed and discussed jointly with the Directorate General of Mines Safety (DGMS), the Chief Mining Advisor to the Railway Board (CMA), Central Mining Research Institute (CMRI) and Bharat Coking Coal Limited (BCCL).

The railway line was made to gradually subside by a maximum of about 800 mm without any disturbance in its operation during the entire period of about 25 years.

II. Salanpur Colliery of Bhrat Coking Coal Limited in Jharia Coalfield

At Salanpur colliery of BCCL in JCF 9.0 m thick X seam was extracted by partial extraction method of widening and heightening of the galleries with hydraulic sand stowing in a panel below Katras railway yard. The method of extraction in the panel was designed such that the factor of safety of the reduced pillars to be left in the stowed goaf was about 1.0 so that there would be practically no subsidence in the railway yard.

The extraction was completed without causing practically any subsidence on the surface in the railway yard.

III. Jamadoba 6 & 7 pits Colliery of Tata Iron & Steel Co. Ltd. in Jharia Coalfield

Extraction of XIV and XII seams was planned by longwall system with stowing below Bhaga Railway Station and yard. The railway lines were made to gradually subside by a maximum of about 300 mm without any interruption in operation over a period of about ten years. Some cracks developed in the railway platform, which could be easily repaired.

IV. Jitpur Colliery of Indian Iron & Steel Co. Ltd. in Jharia Coalfield

A large number of studies were conducted for the extraction of coal seams from underneath and in the vicinity of surface properties. For the first time in the country the concept of extraction alternately with stowing and caving was introduced at this mine.

The most recent study was in respect of the feasibility of extraction of 17 and 18 seams from underneath and in the vicinity of a railway siding by bord and pillar system with hydraulic stowing. Extraction in the seams has been permitted on the basis of the recommendations made after anticipation of the subsidence movements and their impacts. The details of this case study are outlined in *Annexure X*.

V. RK7 & RK New Tech Inclines

Feasibility of extraction of panels in six seams in descending order by bord and pillar system with caving below forest land was studied by anticipating subsidence movements and their impacts. *Annexure XI* gives the details of the panels, subsidence related parameters, anticipated subsidence movements, impacts of subsidence movements, impact mitigation, and recommendations made.

VI. Naheriya Mine

Feasibility of extraction of panels in two seams by bord and pillar system with caving below forest land was studied by anticipating subsidence movements. The details of the panels, subsidence related parameters, predicted subsidence movements, etc. are given in *Annexure XII*.

Anticipated surface subsidence contours due to extraction of both the seams are shown in Fig. 2.23 and the post subsidence surface contours are shown in Fig. 2.24.

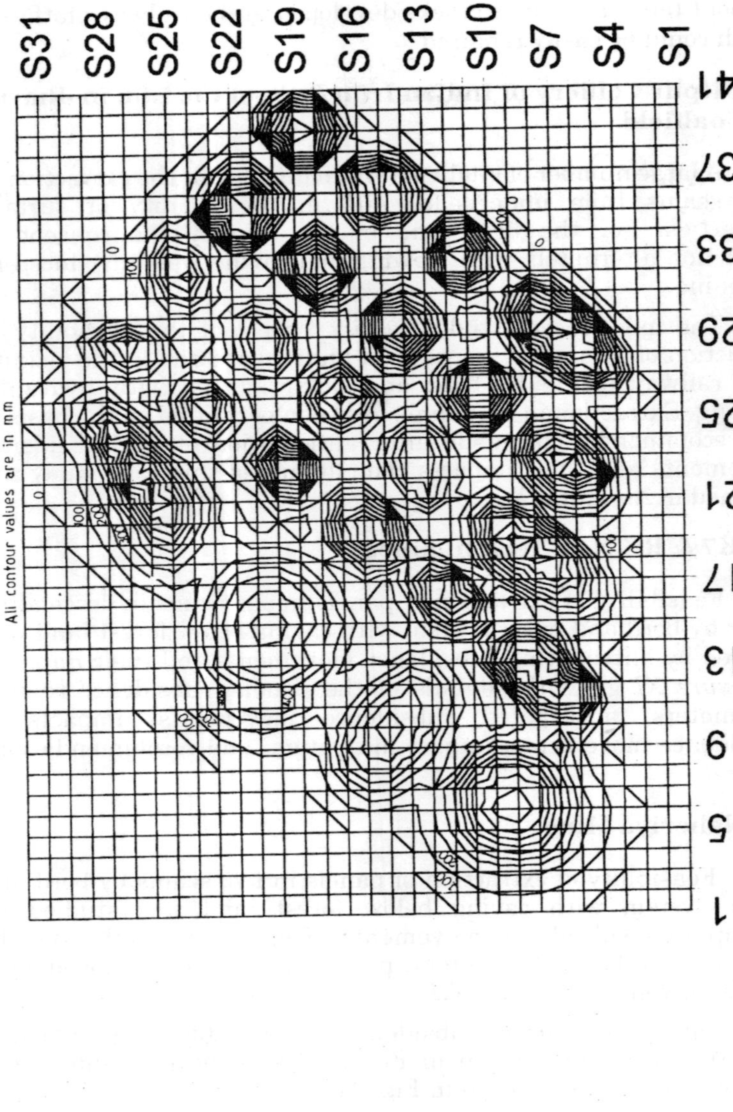

Fig. 2.23. Anticipated subsidence contours (Naheriya mine)

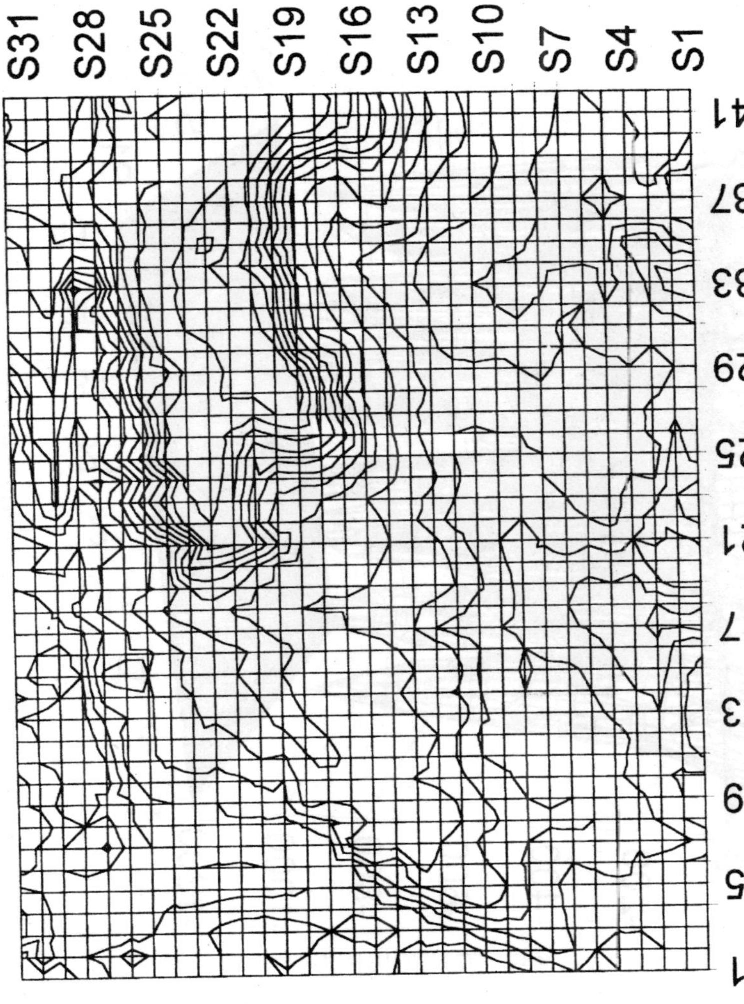

Fig. 2.24. Anticipated post subsidence surface contours (Naheriya mine)

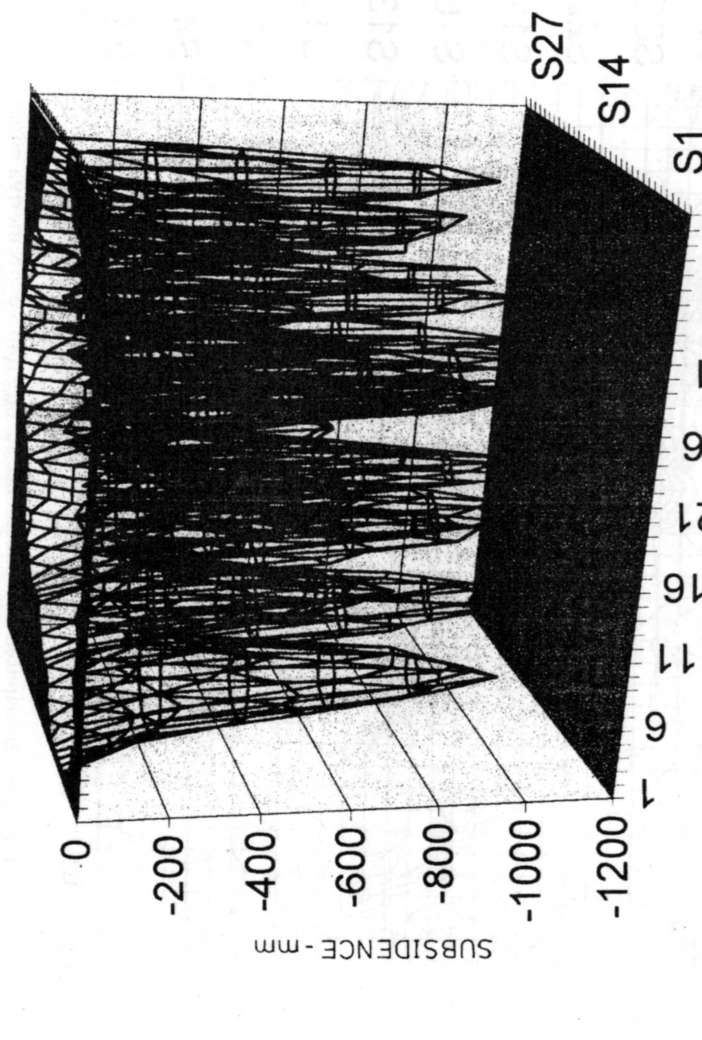

Fig. 2.25. Three dimensional anticipated subsidence (Naheriya mine)

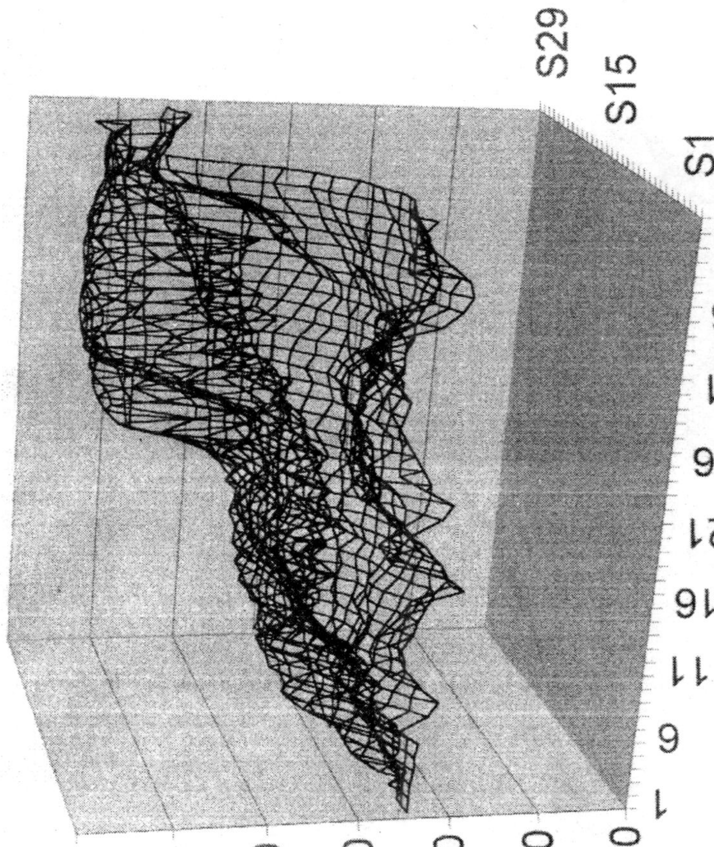

Fig. 2.26. Original surface topography (Naheriya mine)

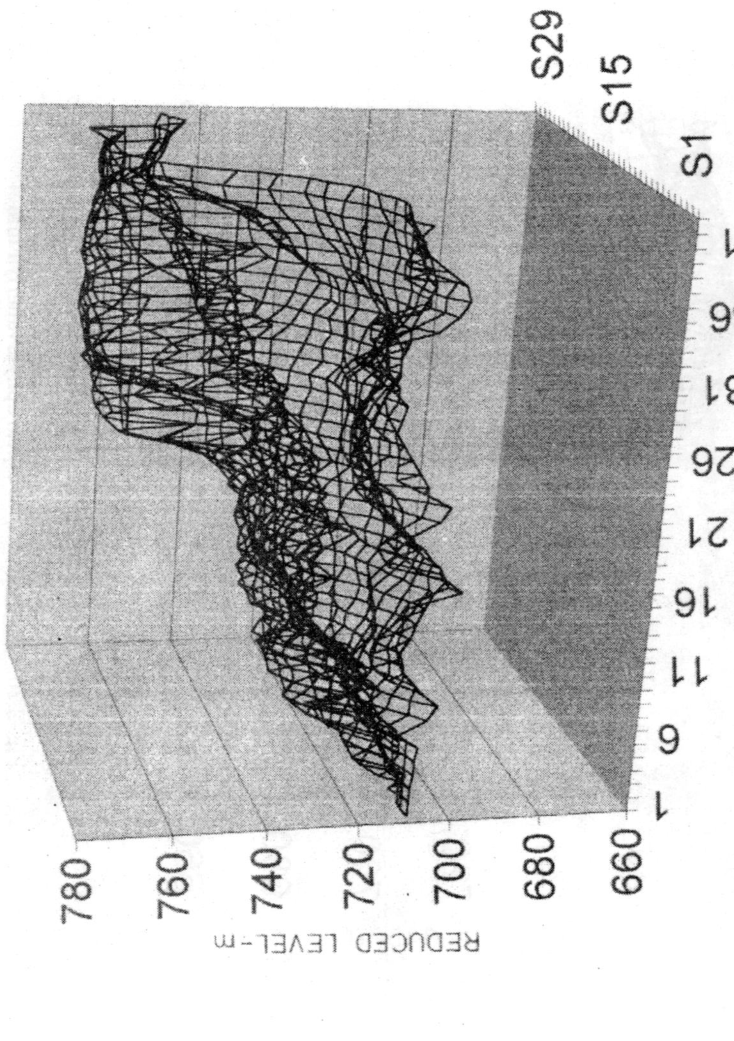

Fig. 2.27. Post subsidence topography (Naheriya Mine)

The three dimensional anticipated subsidence movements due to extraction of both the seams are shown in Fig. 2.25.

The original and anticipated post subsidence surface topography are shown in Fig. 2.26 and 2.27 respectively.

2. Concluding Remarks

It is evident from the above case studies that nearly all aspects of coal mining land subsidence can be predicted to a reasonable degree of accuracy and accordingly the management and mitigative measures can be planned. In this connection it is being suggested that depending upon the geo-mining conditions of mines adequate prediction of the subsidence movements, their impacts and management and mitigative measures should be incorporated in the project report. Wherever necessary the layout of the underground extractions may be redesigned to minimize the impacts of subsidence movements.

It is also suggested that in the mines where subsidence movements are anticipated to have visible impacts on surface properties these movements and their impacts should be monitored.

In the situations where the monitored subsidence movements and their impacts are not in agreement with the predictions, the layout of underground extraction is changed, the sequence of extraction is changed, or the method of working is changed, a fresh subsidence study should be conducted.

ANNEXURE — I

Subsidence Impact Matrix

For the assessment of the impacts of the subsidence movements on the various categories of surface, sub-surface and underground properties the author developed a Subsidence Impact Matrix. To use this matrix the subsidence movements are anticipated and then corresponding to the anticipated movements the impacts on the properties are read on the left hand side. The matrix is as given below.

Sl. No.	Impact	Subsidence mm	Slope mm/m	Strain mm/m
SURFACE TOPOGRAPHY				
1	Practically no impact	<500	<3	<3
2	Some fine cracks or one or two 50 mm wide cracks with visible depression	<500 500-1000	3-5 5-10	3-5 3-5
3	A large number of fine cracks or a few 100 mm wide cracks with marked depression	500-2000	10-20	5-10
4	A large number of 50-100 mm wide cracks or a few 200 mm wide cracks with stepping. Marked distortion in surface topography	500-2000	>20	10-20
5	Up to 500 mm wide cracks with stepping and prominent distortion in surface topography	>1000	>50	20-50
6	Many 500 mm wide cracks, some up to 1000 mm width with large stepping. Severe distortions in surface topography	>2000	>100	50-100
7	Very severe distortions in surface topography. Stepped subsidence with very wide cracks	>2000	>100	>100
SURFACE WATER BODIES (Ponds, Rivers, Nallahs, Jores, HFL)				
1	Practically no impact. No loss of water	<500	<3	<3
2	Marginal impact in some cases only. Some loss of water and water logging	<1000	<5	<5
3	Severe impacts. Major loss of water. Severe water logging	>1000	>5	>5
SUB-SURFACE WATER TABLE				
1	Marginal depletion in water retaining capacity	<500	<3	<3
2	Severe depletion in water retaining capacity	–	–	>5
AQUIFERS				
1	Depletion in water retaining capacity	–	–	>3
WATER LOGGING ON SURFACE				
1	Very little water logging	<500	–	–
2	Some 300-500 mm deep water logging depending on surface topography	500-1000	–	–

3	Marked water logging	>1000	–	–
ROADS				
1	Practically no impact	<500	<5	–
2	Depressions with gentle slope	–	5-10	–
3	Steeper slopes (speed restriction may be necessary)	–	20-50	–
4	Marginal repairs necessary	–	20-50	>10
5	Major repairs necessary	–	>50	>10
RAILWAY LINES - JOINTED CONSTRUCTION				
1	Practically no impact	–	<10	<3
2	Minor to severe impacts, repairs necessary due to bending, twisting and breaking of rails and steeper gradients	–	>10	>3
RAILWAY LINES - WELDED CONSTRUCTION				
1	No subsidence permitted. Even very small strain can cause twisting and breaking of rails			
RAILWAY SIDINGS (Jointed Construction)				
1	Practically no impacts	–	<10	<3
2	Minor to severe impacts, repairs necessary	–	>10	>3
SINGLE STORY HUTMENTS (Kuccha)				
1	Practically no impact. A few fine cracks in plastered walls	–	<5	<3
2	Minor repairable impacts. Fine cracks. A few 10 mm wide cracks.	–	<10	3-5
3	Major/severe impacts, Wide cracks, stepping, tilting etc.	–	>10	>5
SINGLE STORY BUILDINGS				
1.	Very little impact. A few fine cracks or one/two 5-10 mm wide cracks in plaster	–	<5	<3
2	Minor impacts, repairable. 5-10 mm wide cracks, doors and windows getting slight jamming, slight tilting, etc.	–	5-10	3-5
3	Severe impacts, major repairs necessary. Wider cracks, stepping, crushing and marked tilting	–	>10	>5
DOUBLE STORY BUILDINGS				
1.	Very little impact. A few fine cracks or one/two 5-10 mm wide cracks	–	<5	<3
2	Little repairable impact. 5-10 mm wide cracks, slight displacement of walls against roof, doors and windows getting slightly jammed.	–	5-10	3-5
3.	Severe impacts, major repair necessary. Wider Wider cracks, stepping, crushing and tilting. Gaps between walls and roof.	–	>10	>5
MULTI-STORY BUILDINGS				
1	Little impacts, repairable. 5-10 mm wide cracks, doors and windows getting slight jamming, displacement of walls against roof	–	<5	<3

2	Severe impacts. Wider cracks, crushing, tilting and stepping	–	>5	>3
LARGE BUILDINGS, MONUMENTS, HISTORICAL BUILDINGS, ETC.				
1	Very little impacts. A few fine cracks or one/ two 5-10 mm wide cracks.	–	<3	<1.5
2	Little impacts. 5-10 mm wide cracks. Damage to decorations. Slight displacements. Doors and windows getting jammed.	–	3-5	1.5-3
3	Severe impacts. Wider cracks, tilting, crushing, etc. Major repairs necessary.	–	>5	>3
AERIAL ROPEWAYS				
1	Practically no impact	–	<5	<3
2	Little repairable impacts	–	5-10	3-5
3	Severe impacts. Ropes may leave pulleys due to change in alignment. Tilting of pylons. Buckling of structure	–	>10	>5
HIGH TENSION PYLONS				
1	Practically no impact	–	<5	<3
2	Severe impacts. Tilting, buckling and may be collapse of pylons	–	>5	>3
UNDERGROUND CABLES				
1	Practically no impact.	–	–	<3
2	Severe impacts (cables may break due to tension)	–	–	>3
UNDERGROUND PIPELINES				
1	Practically no impact	–	–	<1.5
2	Severe impacts. Breaking of pipes	–	–	>1.5
OVERLYING VIRGIN SEAMS				
1	Practically no impact. No visible signs of subsidence when the seams are developed.	–	<5	0<3
2	A little impact. A little crushing of coal, roof and floor rock. Fire risk when development is done in upper seam.	–	5-10	3-5
3	Severe impacts. Crushing of coal, roof and floor rock. Stepping in tensile strain zone. Fire risk. Heaving. Supports necessary during development.	–	10-20	5-10
4	Very severe impacts. Severe crushing, large stepping and entry into subsided area rather difficult. High fire risk. Arching necessary	–	>20	>10
OVERLYING WORKINGS (Standing on developed pillars)				
1	Practically no impact on galleries and pillars. Some spalling.	–	<5	<3
2	Visible floor lifting, side spalling and roof falls. Supports required. Fire risk	–	–	3-5
3	Marked floor lifting, side spalling and roof falls. High fire risk. Workings unsafe.	–	–	5-10

4	Severe floor lifting, large roof falls, wide spread side spilling, stepping, very high fire risk. Workings unsafe	–	–	>10
OVERLYING WORKINGS (Standing on reduced pillars)				
1	Practically no impact.	–	<3	<1.5
2	Marginal impact on stability of stooks. The stooks with marginal factor of safety may collapse increasing loading on adjoining stooks	–	<5	<5
3	Failure of a few stooks may lead to chain of failures causing partial or total collapse of area, thus causing additional subsidence at the overlying horizons.	–	<5	>5
OVERLYING WORKINGS (Packed or stowed)				
1	Practically no impact on stowed workings. Some loss of water	–	–	<3
2	Complete loss of water from stowed areas and also from adjoining rise side areas	–	–	>3
WATERLOGGED OVERLYING WORKINGS (Standing on developed pillars)				
1	Practically no impact on pillars, galleries and water retaining capacity of the workings	–	–	<3
2	Marginal loss of water through fine cracks in strata around. Dewatered areas may have risk of fires, roof falls, side spilling, floor lifting, etc.	–	–	<5
3	Major loss of water. High fire risk in dewatered areas with roof falls, side spalling, floor lifting, etc.	–	–	5-10
4	Total loss of water. Very high fire risk with severe floor lifting, roof falls and spalling.	–	–	>10
WATERLOGGED OVERLYING WORKINGS (Standing on reduced pillars/stooks)				
1	Stooks may collapse causing additional subsidence on surface	–	–	<3
2	Partial loss of water. Collapse of stooks. Additional surface subsidence. Fire risk	–	–	3-5
3	Total loss of water. Collapse of workings. Additional surface subsidence. Fire risk	–	–	>5
IN CURRENT WORKINGS FROM SUBSIDENCE AT THE LEVEL OF SURFACE				
1	Practically no impact	–	–	<5
2	Leakage of air. Fire in goaves at shallow depths.	–	–	>5
IN CURRENT WORKINGS FROM SUBSIDENCE AT THE LEVEL OF OVERLYING WATER BODIES				
1	Practically no impact	–	–	<3
2	Marginal increase in make of water	–	–	3-5
3	Appreciable increase in make of water	–	–	5-10
4	Heavy increase in make of water, which may lead to inundation	–	–	>10

SURFACE ATMOSPHERE				
1	Practically no impact.	–	–	<5
2	Some air from underground workings at shallow depths may leak to surface	–	–	5-10
3	Leakage from shallow depth workings. If the workings have fire, surface atmosphere is likely to be polluted by gases coming from the fire	–	–	>10
SUB-SOIL				
1	Practically no impact	–	–	<3
2	Very little impact in the form of reduction of water retaining capacity	–	–	3-5
3	Temporary loss in water retaining capacity. Cracks filling may improve water retaining capacity	–	–	5-10
4	Long term loss of water retaining capacity. Suitable protective measures necessary	–	–	>10
AGRICULTURE				
1	Practically no impact	–	–	<5
2	Marginal impact, i.e., reduction in yield due to loss in water retaining capacity of sub-soil	–	–	5-10
3	Major impact, i.e., sizeable reduction in yield	–	–	>10
FOREST AND PLANTATION				
1	Practically no impact	–	<10	<5
2	Temporary loss in water retaining capacity of top-soil may affect undergrowth slightly. Slight tilting of plants/trees	–	10-20	5-10
3	Short term impact on trees in zones having cracks. The cracks may get filled in due course. Tilting of trees	–	20-50	10-20
4	Wide cracks may severely affect undergrowth but may not have much impact on large trees except those in the tensile strain zone where wide cracks develop. High tilting may cause some trees to fall in the high slope zone	–	>50	>20

ANNEXURE – II

Controlling Subsidence Movements by Reducing Extraction Thickness

Subsidence movements on the surface or any plane of reference above the coal seams under extraction can be controlled by controlling the extraction thickness. This is illustrated by an example given hereunder.

A 3.5 m thick coal seam is to be extracted by longwall system with caving. The extraction details are as given below.

Size of the panels : 150 m x 1500 m

Depth of the panels : 100 m

Overlying rock mass : Inter-bedded shales and sandstones

Surface properties : Forest area

Hence, the subsidence related parameters are as outlined below.

Rock mass factor (M) : 0.6

Critical width-depth ratio : 1.4 times the depth

Width-depth ratio of panels : 1.5 (Hence all the panels are of super-critical width-depth ratio)

Extraction thickness (m′) : 3.5 m

Depth factor (h′) : 1.0

Extraction percentage factor (e′): 0.95

Goaf treatment factor (a′) : 0.95

Therefore, the maximum anticipated subsidence movements are as computed below.

$S_{max} = 500\ (1 + 0.6)(0.95 \times 0.95 \times 1.0 \times 3.5) = 2527$ mm

$G_{max} = 2 \times 2527 / 100 = 50.5$ mm/m

$E_{max(-)} = 1.0 \times 2527 / 100 = 25.3$ mm/m

$E_{max(+)} = 1.05 \times 2527 / 100 = 26.6$ mm/m

The maximum permissible tensile strain in the forest areas is 20 mm/m. Hence, the maximum anticipated tensile strain in this case is more than the permissible limit. This can be reduced by reducing the thickness of extraction of the seam as seen from the anticipated maximum subsidence movements for the different extraction thicknesses given below.

Extraction thickness	Smax	Gmax	Emax(-)	Emax(+)
m	mm	mm/m	mm/m	mm/m
3.25	2347	46.9	23.5	24.7
3.00	2166	43.3	21.7	22.8
2.75	1986	39.7	19.9	20.9
2.50	1805	36.1	18.1	19.0

Hence, in this case the extraction thickness in the panels should be restricted to 2.5 m.

ANNEXURE – III

Controlling Subsidence Movements by Optimizing the Width-Depth Ratio

Subsidence movements can be controlled by optimizing the width-depth ratio of the panels. This is illustrated with the help of an example in this annexure.

A 3.0 m thick seam is proposed to be extracted below surface water bodies and a railway siding of jointed construction. Other details of the extraction are as given below.

Depth of the workings : 120 m

Method of working : Bord and pillar

Goaf support : Caving

Percentage of extraction in panel : 75

Overlying rock mass : Inter-bedded shales and sandstones

Thus, the subsidence related parameters are as outlined below.

Rock mass factor (M) : 0.6

Depth factor (h′) : 1.0

Goaf treatment factor (a′) : 0.95

Extraction percentage factor (e′) : 0.5

Extraction thickness (m′) : 3.0 m

The safe limits of the subsidence movements for the surface properties are as given below

1 For surface water bodies $E_{(+)}$ = 3.0 mm/m

2 For railway lines of jointed construction E = 3.0 mm/m

The anticipated maximum possible subsidence movements are as computed below.

Smax = 500 (1+0.6)(0.5 x 0.95 x 1.0 x 3.0) = 1140 mm

Gmax = 2 x 1140 / 120 = 19.0 mm/m

$Emax_{(-)}$ = 1.0 x 1140 / 120 = 9.5 mm/m

$Emax_{(+)}$ = 1.05 x 1140 / 120 = 10.0 mm/m

Since the anticipated maximum possible subsidence movements are more than the permissible limits the panels are to be designed with sub-critical width-depth ratio so as to bring the anticipated maximum subsidence movements within the permissible limits.

The anticipated maximum subsidence movements for the different sub-critical width-depth ratios are as given below.

Width-depth ratio	S	G	$E_{(-)}$	E(+)
	mm	mm/m	mm/m	mm/m
1.4	1140	19.0	9.5	10.0
1.2	1026	18.0	9.0	9.5
1.0	855	17.8	8.9	9.3
0.8	513	12.9	6.45	6.8
0.6	171	5.7	2.85	3.0

Therefore, in the present situation the width of the underground extractions should be planned so as not to exceed 0.6 times the depth. Since, the depth is 120 m, the width of extraction to be planned is 72 m.

ANNEXURE – IV

Controlling Subsidence Movements by Planning Extraction With Hydraulic Stowing

Subsidence movements can also be controlled and greatly reduced by planning underground extractions with hydraulic stowing. This is explained with the help of an example hereunder.

In the case study in Annexure - II, if the extraction is to planned with hydraulic stowing the extraction and subsidence related parameters will be as given below.

Size of panels : 150 m x 1500 m

Extraction thickness : 3.5 m

Depth : 100 m

Width-depth ratio : 1.5

Rock mass factor : 0.6

Extraction percentage factor : 0.95

Depth factor : 1.0

Goaf treatment factor : 0.1

Hence, the anticipated maximum possible subsidence movements are as computed below.

$$Smax = 500 (1 + 0.6)(0.95 \times 0.1 \times 1.0 \times 3.5) = 266 \text{ mm}$$

$$Gmax = 2 \times 266 / 100 = 5.3 \text{ mm/m}$$

$$Emax_{(-)} = 1.0 \times 266 / 100 = 2.7 \text{ mm/m}$$

$$Emax_{(+)} = 1.05 \times 266 / 100 = 2.8 \text{ mm/m}$$

Thus, with hydraulic stowing the anticipated maximum subsidence movements have greatly reduced and are well within the permissible limits for the surface water bodies and the jointed construction railway lines.

ANNEXURE – V

Controlling Subsidence Movements by Planning Extraction Without Barriers and by Alternate Stowing and Caving

The shape of the subsidence trough and thereby the curvature due to subsidence can be controlled by planning the underground workings in various ways. The most important amongst them are barrier-less mining and alternate stowing and caving. In case of barrier-less mining the subsidence trough in the final stage has curvature on the two ends only. Extraction with alternate stowing and caving is planned invariably for the protection of the surface properties such that the areas below the properties are extracted with stowing while the ones not having surface properties are extracted with caving. These are illustrated with the help of an example each below.

Extraction Without Barrier With Caving

A 3.0 m thick coal seam is to be extracted by longwall system with caving at a depth of 300 m with panel sizes of 150 m x 1500 m without leaving any barriers in between. The overlying rock mass consists of inter-bedded sandstones and shales.

Barrier-less mining can be designed in two sequences as shown in Fig. 2.12. In the first sequence the width-depth ratio increases with every panel while in the second sequence it increases when the intermediate panels are extracted. The maximum subsidence movement prediction for the two sequences is given below.

Sequence 1 : Successive extraction of the panels.

In this example the subsidence related parameters are as outlined hereunder.

M = 0.6; m′ = 3.0 m; w = 150 m; h = 300 m;

h′ = 1.1; e′ = 0.95; and a′ = 0.95

The anticipated maximum subsidence movements are as given below.

Panels	Width of extraction	Width-depth ratio	S	G	E(–)	E(+)
	m		mm	mm/m	mm/m	mm/m
1	150	0.5	119	2.0	1.0	1.0
1 & 2	300	1.0	1787	14.9	7.5	7.9
1, 2 & 3	450	1.5	2383	15.9	8.0	8.4
1, 2, 3, & 4	600	2.0	2383	15.9	8.0	8.4

Sequence 2 : Alternate extraction of the panels.

In this sequence the subsidence related parameters are the same as in the first sequence and the anticipated subsidence maximum movements are as given below.

Panels	Width of extraction	Width depth ratio	S	G	E(–)	E(+)
	m		mm	mm/m	mm/m	mm/m
1	150	0.5	119	2.0	1.0	1.0
3	150	0.5	119	2.0	1.0	1.0
1, 2 & 3	450	1.5	2383	15.9	8.0	8.4
5	150	0.5	119	2.0	1.0	1.0
1, 2, 3, 4 & 5	750	2.5	2383	15.9	8.0	8.4

The anticipated subsidence profiles in these two sequences are graphically presented in Fig. 2.28 and 2.29 respectively.

Extraction with Alternate Stowing and Caving

In the two sequences discussed above in this annexure if the extraction is planned alternately with caving and stowing the subsidence movements will greatly reduce because the

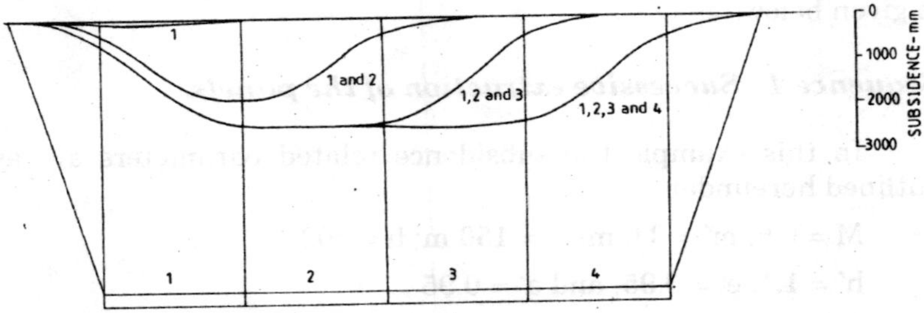

Fig. 2.28. Anticipated subsidence profiles (Sequence 1)

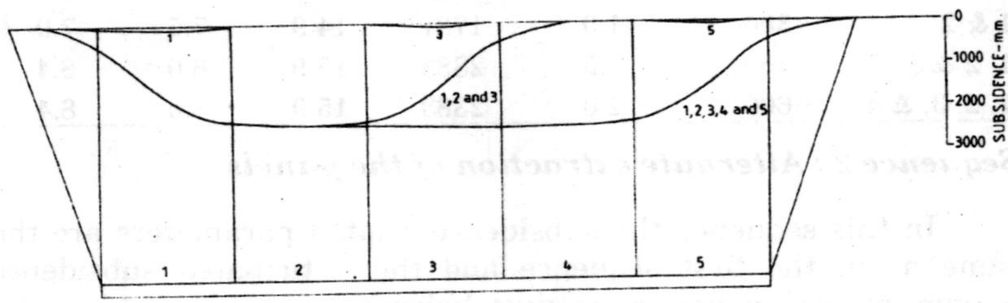

Fig. 2.29. Anticipated subsidence profiles (Sequence 2)

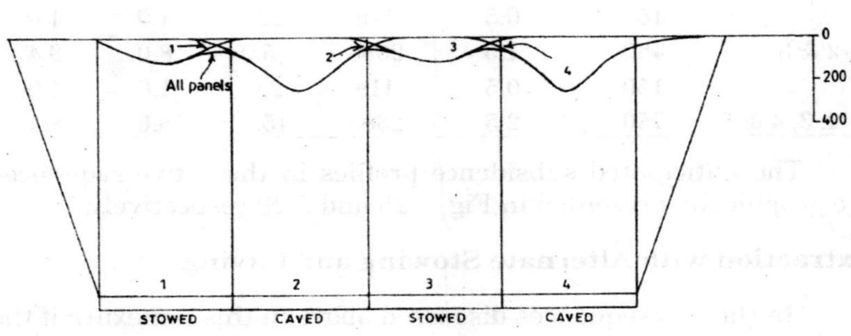

Fig. 2.30. Anticipated subsidence profiles for extraction with alternate stowing and caving.

hydraulically stowed goaves will act like yielding barriers. In such a situation the anticipated subsidence movements will be as given below and subsidence profiles will be as shown plotted in Fig. 2.30.

Panels	Width of extraction	Width-depth ratio	S	G	E(–)	E(+)
	m		mm	mm/m	mm/m	mm/m
1	150	0.5	119	2.0	1.0	1.0
3	150	0.5	119	2.0	1.0	1.0
2	450	1.5	251	1.7	0.9	0.9
4	600	2.0	251	1.7	0.9	0.9

Thus, the subsidence movements will be much less. In this sequence the width of extraction during the extraction of the third panel with hydraulic stowing will behave as that of the three panels and the angle of draw will be about 40 - 45^{0} on the goaf sides of the stowed panels.

ANNEXURE – VI

Controlling Subsidence Movements by Suitably Locating the Barriers

The shape of the subsidence profiles can be controlled by suitably locating the barriers in multiple seam mining. This is illustrated by the following example.

Two 3.0 m thick seams are to be planned on panel and barrier system. The extraction details are as given below.

Depth of upper seam : 120 m

Parting between the seams : 30 m

Panel width in both the seams : 150 m

Method of working : Bord & pillar caving

Width of barriers : 25 m (solid)

Overlying rock mass : Inter-bedded shales & sandstones

Extraction percentage in panels : 75

Hence, subsidence related parameters are as given below.

For upper seam w/h = 1.25; M = 0.6; m′ = 3.0 m; e′ = 0.5; a′ = 0.95; & h′ = 1.0.

For lower seam w/h = 1.0; M = 0.8; m′ = 3.0 m; e′ = 0.5; a′ = 0.95; & h′ = 1.0.

Therefore, the maximum subsidence movements for the individual panels will be as outlined hereunder.

For upper seam S = 1054 mm; G = 18.0 mm/m; $E_{(-)}$ = 9.0 mm/m; & $E_{(+)}$ = 9.5 mm/m.

For lower seam S = 1154 mm; G = 16.2 mm/m; $E_{(-)}$ = 8.1 mm/m; & $E_{(+)}$ = 8.5 mm/m.

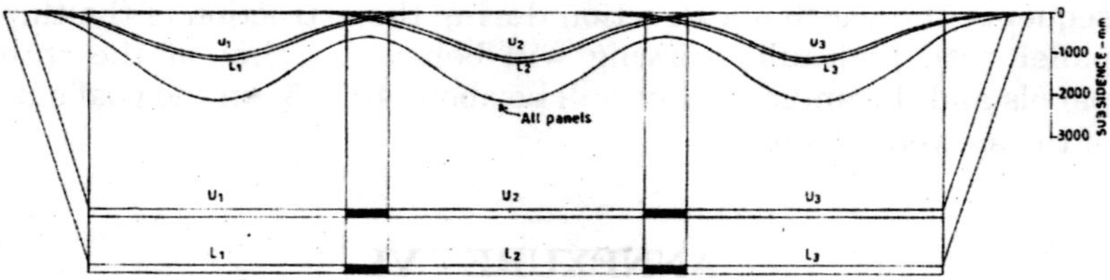

Fig. 2.31. Anticipated subsidence profile for superimposed extraction

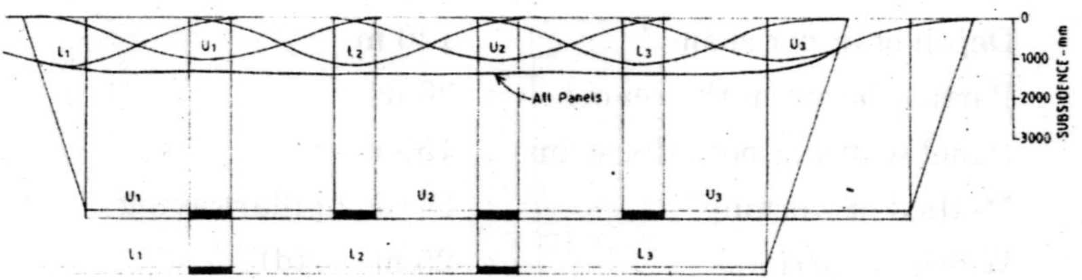

Fig. 2.32. Anticipated subsidence profile for staggered extraction

While planning for the extraction of the seams the barriers can either be kept in superimposition or staggered. Subsidence profiles resulting in these two conditions of planning are shown in Fig. 2.31 and 2.32. It can be seen that in the staggered planning the curvature of the subsidence trough is less and hence the impacts on the surface will also be less.

ANNEXURE – VII

Extraction Planning with NEW Concept

In the situations where practically no subsidence movement is desired on the surface or the plane of reference above the seam under extraction underground extraction planning is done on the concept of non-effective width-depth ratio. This is explained below with the help of an example.

A coal seam of about 3.0 m thickness is to be planned for extraction at a depth of about 200 m by bord and pillar system with caving in such a manner that there is practically no subsidence on the surface. The overlying rock mass consists of predominantly sandstones. How the extraction planning shall be done in this situation?

In this example, – rock mass factor = 0.5

– depth = 200 m

Hence, NEW = 1 - M = 1 - 0.5 = 0.5

Therefore, the width of extraction that should be taken into consideration for planning underground extractions is 200 x 0.5 = 100 m. In such a planning the barriers between the extractions should be of adequate width with a factor of safety of 1.0 and more for hydraulic stowing and 2.0 and more for caving.

ANNEXURE – VIII

Determination of the Factor of Safety of the Barrier Pillars, and Hence their Designing

In partial extraction designing, either by bord and pillar or longwall method, it becomes necessary to design barriers to be left between the panels to be extracted. In order to meet the design requirement the barriers should be stable from long-term consideration with a factor of safety >1.0 for extractions with hydraulic stowing and >2.0 for extractions with caving. For determining the factor of safety of the pillar and hence their designing the following methodology is used.

Determination of the strength of the barrier pillars with the help of the following equation (Anon, 1987 and Shiorey & Singh, 1989)

$$S = 0.27\ \sigma_c\ h^{-0.36} + (H/250 + 1)(We/h - 1)$$

Where, S = strength of the pillars in MPa,

σ_c = strength of 2.5 cm cube of coal in the seam in MPa,

h = working height or height of the pillar in m,

H = maximum depth of the workings in the panel in m, and

We = equivalent pillar width in m (this is equal to 4 x pillar area/pillar perimeter).

Determination of load on the barrier pillars using the following or any other suitable equation (Salamon, 1967)

$$P = 0.025\ H\ (w + 0.3H)(w_1 + B)/w.w_1$$

When the extraction width is greater than 0.6H, and when it is less the average load is given by the following equation

$$P = 0.025\ [H(w + L) - L^2/1.2]\ (w_1 + B)/w.w_1$$

Where, P = load on the pillar in MPa,

w = width of the pillar corner to corner in m,

w_1 = length of the pillar corner to corner in m,

L = width of extraction in m, and

B = gallery width in m.

An example of the determination of the factor of safety of the barrier pillars for designing the underground workings at Balgi Project of South Eastern Coalfields Limited is given below.

S.N.	Panel	B	w	H*	σ_c	h	We	Factor of safety
		m	m	m	MPa	m	m	
1	E4B	4.2	20.6	116	41	3.1	20.6	2.11
2	E4A	4.2	20.6	133	41	3.2	20.6	1.81
3	E4	4.2	22.2	166	41	3.5	22.2	1.49
4	E5	4.2	23.8	189	41	4.0	23.8	1.28
5	E6	4.2	25.3	216	41	3.9	25.3	1.25
6	E7	4.2	25.3	235	41	3.9	25.3	1.09

*Maximum depth of the panels

ANNEXURE – IX

Computation of the Volume of Subsidence Trough

Knowledge of the volume of the subsidence trough is important from the point of view of the management of subsided land as this helps in the assessment of the quantity of the water that can be stored in the depressions caused by subsidences and also in the assessment of the quantity of solids required for restoration/reclamation of the surface topography. An example of computation of the volume of the subsidence trough is given hereunder.

The volume of the subsidence trough is computed with the help of the subsidence contours which are plotted on the basis of the predictions or the observations. The areas enclosed by all the contours are computed as A_1, A_2, A_3,.................A_n. The volume of the trough is computed with the help of the area of the contours, contour values and their interval as outlined below.

Contour line	Value of contour	Area of contour line	Contour interval	Volume between contour lines
	m	m^2	m	m^3
A_1	0			
			0.1	2,925
A_2	0.1	28,000		
			0.1	2,600
A_3	0.2	24,000		
			0.1	1,950
A_4	0.3	15,000		
			0.1	1,250
A_5	0.4	10,000		
			0.1	750
A_6	0.5	5,000		
			0.1	300
A_7	0.6	1,000		
			0.1	50.5
A_8	0.7	10		
			Total volume	**9,825.5**

ANNEXURE – X

Details of Jitpur Colliery's Case Study

Feasibility of extraction of seven panels, three in 17 seam and four in 18 seam with hydraulic sand stowing was studied. The details of the panels are as given below.

1 Coalfield : Jharia

2 Colliery : Noonidih-Jitpur

3 Company : Indian Iron & Steel Co. (IISCO)

4 Coal seams : 17 & 18

5 Seam thickness : 2.44 m & 3.37 m respectively

6 Extraction thickness : 2.2 m & 3.3 m respectively

7 Average size, depth and status of panels

Seam	Panel	Size	Depth	Status
		m x m	m	
17	P17A	290x145	130-173	Partially extracted with stowing & partially developed on bord and pillar pattern
17	P17B	680x140	94-150	Developed on bord and pillar pattern
17	P17C	500x140	20-86	Developed on bord and pillar pattern
18	P1(A)	225x280	72-112	Extracted with stowing
18	P2	92x125	16-34	Having developed pillars of various sizes
18	P3	70x360	60-93	Standing on developed pillars
18	P4	200x430	15-55	Standing on developed pillars

8 Overlying strata : Inter-bedded sandstones and shales

9 Method of working : Longwall and bord and pillar with hydraulic sand stowing

10 Overlying workings : No overlying workings over the panels in 18 seam. However the panels in 17 seam had goaves and developed workings of 18 seam above them

11 Surface properties

Seam	Panel	Surface properties
17	P17A	Main office; IISCO quarters; Seasonal drains; Private dwellings
17	P17B	Pump house; Railway siding; IISCO bunglows
17	P17C	Noonidih pit; Overman colony; Sand bunker; Quarters; Railway siding; Private dwellings, Aerial ropeway
18	P2	A school building; A village road
18	P3	JBO loop siding; Colliery buildings: A road
18	P4	IISCO buildings & road; Hutments; A drain

12 Details of railway siding : Jointed construction with a long gradient of 1 in 335

13 Extractable reserves : 10,72,500 tonne

14 Subsidence related parameters of the panels

Seam	Panel	m′ m	h (avg) m	w/h	M	e′	a′	h′
17	P17A	2.2	151.5	0.96	0.8	0.95/0.6	0.07-0.1	1.0
17	P17B	2.2	122	1.15	0.8	0.95/0.6	0.07-0.1	1.0
17	P17C	2.2	53	2.64	0.8	0.95/0.6	0.07-0.1	1.0
18	P1(A)	3.3	93.5	2.41	0.8	0.95/0.6	0.07-0.1	1.0
18	P2	3.3	25	3.68	0.8	0.95/0.6	0.07-0.1	1.0
18	P3	3.3	76.5	0.91	0.8	0.95/0.6	0.07-0.1	1.0
18	P4	3.3	35	5.71	0.8	0.95/0.6	0.07-0.1	1.0

15 Safe limits of subsidence movements for the surface properties

S. No.	Property	Safe limit
1.	Buildings and hutments	Strain = 3 mm/m for slight repairable damages
2.	Railway siding of jointed construction	Slope = 10 mm/m (1 in 100) Strain = 3 mm/m
3.	Drains	Tensile strain = 4.5 mm/m

The anticipated maximum possible subsidence movements on the surface over all the panels for the different depths and also for their extraction with longwall and bord and pillar methods are as given below.

Seam	Panel	Method	Depth	S_{max}	G_{max}	$E_{max(-)}$	$E_{max(+)}$
			m	mm	mm/m	mm/m	mm/m
17	All	LW	20	132	13.2	6.6	6.9
			40	132	6.6	3.3	3.5
			60	132	4.4	2.2	2.3
			80	132	3.3	1.7	1.8
			100	132	2.6	1.3	1.4
			120	132	2.2	1.1	1.2
			140	132	1.9	1.0	1.1
			160	132	1.7	0.9	1.0
			180	132	1.5	0.8	0.9
		BP	20	100	10.0	5.0	5.3
			40	100	5.0	2.5	2.6
			60	100	3.3	1.7	1.8
			80	100	2.5	1.3	1.4
			100	100	2.0	1.0	1.1
			120	100	1.7	0.9	1.0
			140	100	1.4	0.7	0.7
			160	100	1.3	0.65	0.7
			180	100	1.1	0.55	0.6
18	P1(A)	LW	75	197.5	5.3	2.6	2.8
			80	197.5	4.9	2.5	2.6
			100	197.5	4.0	2.0	2.1
			120	197.5	3.3	1.6	1.7
	P2-P4	BP	20	148.5	14.9	7.4	7.8
			40	148.5	7.4	3.7	3.9
			60	148.5	5.0	2.5	2.6
			80	148.5	3.7	1.9	2.0
			100	148.5	3.0	1.5	1.6
			120	148.5	2.5	1.2	1.3

LW - Longwall; BP - Bord and Pillar

The feasibility of extraction of each panel was studied while considering the anticipated maximum possible subsidence movements, the presence of the surface properties and the safe limits of the subsidence movements. The following recommendations were made.

1. Wherever the panels in the two seams are in superimposition extraction is to be done in descending order, i.e., 18 seam extraction first followed by 17 seam extraction.

2. Shaft protection pillars of J1, J2 and J3 shafts in 17 seam should be of 95 m radius. Hence, no extraction in this seam should be done within 95 m of these shafts.

3. For N1 and N2 shafts a protection pillar of 60 m radius in 17 seam should be left. Hence, no extraction in this seam should be done within 60 m of these shafts.

4. The panels in 17 seam can be extracted in the following manner.

   Panel P17A - Longwall/bord and pillar with hydraulic stowing

   Panel P17B - Longwall/bord and pillar with hydraulic stowing

   Panel P17C - Longwall/bord and pillar with hydraulic stowing at depths more than 50 m, and partial extraction with hydraulic stowing at depths less than 50 m.

5. The panels in 18 seam can be extracted in the following manner.

   Panel P1(A) - The panel had already been extracted with hydraulic stowing

   Panel P2 - Partial extraction with hydraulic stowing

   Panel P3 - Bord and pillar with hydraulic stowing

   Panel P4 - Partial extraction with hydraulic stowing

6. *The following precautions were suggested.*

   — The panels in the two seams are to be made approachable and their offset plans be made with a fresh survey.

   — The partial extraction methodology with hydraulic stowing wherever applicable as per the recommendations will be decided after knowing the sizes of the pillars and galleries, i.e., after the offset plans are ready.

   — During extraction of all the panels care will be taken that there is no lag in stowing and that practically no gap is left along the roof.

   — Subsidence movements and their impacts on the surface properties shall be monitored over all the panels during and after the extraction as necessary.

   — Damage to the buildings, etc. due to subsidence will be watched and repaired by IISCO.

— Subsidence movements shall be monitored along the railway siding and if necessary maintenance will be done. In this connection IISCO may request the railways to do maintenance of the siding tracks as per the railway schedule.

ANNEXURE - XI

Details of RK7 & RK New Tech Inclines' Case Study

At RK7 and RK New Tech Incline mines of Singareni Collieries Company Limited it was proposed to extract panels in six seams by bord and pillar system with caving below forest land. The study of the feasibility of extraction was done by studying the details of the panels and anticipating the subsidence movements and their impacts on the forest area. The details of the study are given hereunder.

Details of The Panels

The details of the panels under consideration in this study in the six seams as provided by the SCCL were as given below.

1	Mine/colliery	:	RK7 and RK New Tech Inclines
2	Company	:	Singareni Collieries Company Limited (SCCL)
3	Seams	:	1A, 1, 3B, 3A, 3 and 4 Seams
4	Panels	:	The details of the panels under consideration are given later
5	Method of working	:	Bord and pillar
6	Goaf support	:	Caving
7	Anticipated percentage of extraction	:	80
8	Geological disturbances in the panels	:	Practically no major geological disturbances
9	Surface properties	:	Forest land
10	Width of barriers between the panels	:	30 - 40 m
11	Sequence of extraction	:	In descending order and with panels in superimposition
12	Total extractable reserves in the panels	:	About 3.13 million tonne
13.	Parting between the seams	:	10 - 55 m
14	Gradient of the seams	:	1 in 4.5

Details of the Panels in 1A Seam

Sl.No.	Panel	Extraction thickness	Depth	Size of panels	R serves (extractable)	Remarks
		m	m	mxm	tonne	
1	2	3	4	5	6	7
1	Z	3.6	218-248	120X100	64800	Total reserves = 518400 tonne
2	A6	3.6	192-218	130X100	70200	
3	A12	3.6	154-183	130X100	70200	
4	A16	3.6	107-147	180X100	97200	
5	39	3.6	65-100	160X100	86400	
6	33	3.6	35-65	240X100	129600	

Details of the Panels in 1 Seam

Sl.No.	Panel	Extraction thickness	Depth	Size of panels	Reserves (extractable)	Remarks
		m	m	mxm	tonne	
1	2	3	4	5	6	7
1	Z	2.75	238-268	120X100	49500	Total reserves = 396000 tonne
2	A6	2.75	212-238	130X100	53625	
3	A12	2.75	174-203	130X100	53625	
4	A16	2.75	127-167	180X100	74250	
5	39	2.75	85-120	160X100	66000	
6	33	2.75	55-85	240X100	99000	

Details of the Panels in 3B Seam

Sl.No.	Panel	Extraction thickness	Depth	Size of panels	Reserves (extractable)	Remarks
		m	m	mxm	tonne	
1	2	3	4	5	6	7
1	Z	1.83	283-313	120X100	32940	Total reserves = 299205 tonne
2	A6	1.83	257-283	130X100	35685	
3	A12	1.83	219-248	130X100	35685	
4	A16	1.83	172-212	180X100	49410	
5	39	1.83	130-165	160X100	43920	
6	33	1.83	100-130	240X100	65880	
7	34	1.83	64-93	130X100	35685	

Details of the Panels in 3A Seam

Sl.No.	Panel	Extraction thickness	Depth	Size of panels	Reserves (extractable)	Remarks
		m	m	mxm	tonne	
1	2	3	4	5	6	7
1	Z	2.93	293-323	120X100	52740	Total reserves = 637275 tonne
2	A6	2.93	267-293	130X100	57135	
3	A12	2.93	229-258	130X100	57135	
4	A16	2.93	182-222	180X100	79110	
5	39	2.93	140-175	160X100	70320	
6	33	2.93	110-140	240X100	105480	
7	34	2.93	74-103	130x100	57135	
8	42	2.93	81-110	120x100	52740	

Details of the Panels in 3 Seam

Sl.No.	Panel	Extraction thickness	Depth	Size of panels	Reserves (extractable)	Remarks
		m	m	mxm	tonne	
1	2	3	4	5	6	7
1	Z	4.2	348-378	120X100	75600	Total reserves = 789264 tonne
2	A6	4.2	322-348	130X100	81900	
3	A12	4.2	284-313	130X100	81900	
4	A16	4.2	237-277	180X100	113400	
5	39	4.2	195-230	160X100	100800	
6	33	4.2	165-195	240X100	151200	
7	34	4.2	128-158	130x100	81900	
8	42	4.2	136-165	120x100	75600	
9	50	4.2	97-113	120x70	26964	

Details of the Panels in 4 Seam

Sl.No.	Panel	Extraction thickness	Depth	Size of panels	Reserves (extractable)	Remarks
		m	m	mxm	tonne	
1	2	3	4	5	6	7
1	Z	2.14	360-390	120X100	38520	Total reserves = 492414 tonne
2	A6	2.14	334-360	130X100	41730	
3	A12	2.14	296-325	130X100	41730	
4	A16	2.14	249-289	180X100	57780	
5	39	2.14	207-242	160X100	51360	
6	33	2.14	177-207	240X100	77040	
7	34	2.14	141-170	130X100	41730	
8	42	2.14	148-177	120X100	38520	
9	50	2.14	109-125	120X70	26964	

On the basis of the above details the subsidence related parameters of the panels in the six seams were as given below. In these parameters the width-depth ratio was taken with respect to the minimum depth, the extraction percentage factor was taken as 0.6 because the anticipated extraction percentage from the panels was 80, the depth factor was taken as 1.0 for depths up to 250m and 1.1 for depths between 250 and 400m, the goaf treatment factor was taken as 0.95 for extraction with caving, and the rock mass factor was taken as 0.6 for the top most seam and 0.8-0.9 for the lower seams as during extraction of the lower seams the overlying rock mass will be partially disturbed. For the panels in the lower seams having intact overlying rock mass the rock mass factor was taken as 0.6.

Subsidence Related Parameters for Panels in 1A Seam

Sl. No.	Panel	Extraction thickness m	Minimum depth m	With depth ratio	Depth factor	Extraction percentage factor	Goaf treatment factor	Rock mass factor
1	2	3	4	5	6	7	8	9
1	Z	3.6	218	0.46	1.0	0.6	0.95	0.6
2	A6	3.6	192	0.52	1.0	0.6	0.95	0.6
3	A12	3.6	154	0.65	1.0	0.6	0.95	0.6
4	A16	3.6	107	0.93	1.0	0.6	0.95	0.6
5	39	3.6	65	1.54	1.0	0.6	0.95	0.6
6	33	3.6	35	2.86	1.0	0.6	0.95	0.6

Subsidence Related Parameters for Panels in 1 Seam

Sl. No.	Panel	Extraction thickness m	Minimum depth m	With depth ratio	Depth factor	Extraction percentage factor	Goaf treatment factor	Rock mass factor
1	2	3	4	5	6	7	8	9
1	Z	2.75	238	0.42	1.0	0.6	0.95	0.8
2	A6	2.75	212	0.47	1.0	0.6	0.95	0.8
3	A12	2.75	174	0.51	1.0	0.6	0.95	0.8
4	A16	2.75	127	0.79	1.0	0.6	0.95	0.8
5	39	2.75	85	1.18	1.0	0.6	0.95	0.8
6	33	2.75	55	1.82	1.0	0.6	0.95	0.8

Subsidence Related Parameters for Panels in 3B Seam

Sl. No.	Panel	Extraction thickness m	Minimum depth m	With depth ratio	Depth factor	Extraction percentage factor	Goaf treatment factor	Rock mass factor
1	2	3	4	5	6	7	8	9
1	Z	1.83	283	0.35	1.1	0.6	0.95	0.8
2	A6	1.83	257	0.39	1.1	0.6	0.95	0.8
3	A12	1.83	219	0.46	1.0	0.6	0.95	0.8
4	A16	1.83	172	0.58	1.0	0.6	0.95	0.8
5	39	1.83	130	0.78	1.0	0.6	0.95	0.8
6	33	1.83	100	1.00	1.0	0.6	0.95	0.8
7	34	1.83	64	1.56	1.0	0.6	0.95	0.6

Subsidence Related Parameters for Panels in 3A Seam

Sl. No.	Panel	Extraction thickness m	Minimum depth m	With depth ratio	Depth factor	Extraction percentage factor	Goaf treatment factor	Rock mass factor
1	2	3	4	5	6	7	8	9
1	Z	2.93	293	0.34	1.1	0.6	0.95	0.9
2	A6	2.93	267	0.37	1.1	0.6	0.95	0.9
3	A12	2.93	229	0.44	1.0	0.6	0.95	0.9
4	A16	2.93	180	0.56	1.0	0.6	0.95	0.9
5	39	2.93	140	0.71	1.0	0.6	0.95	0.9
6	33	2.93	110	0.91	1.0	0.6	0.95	0.9
7	34	2.93	74	1.35	1.0	0.6	0.95	0.9
8	42	2.93	81	1.23	1.0	0.6	0.95	0.6

Subsidence Related Parameters for Panels in 3 Seam

Sl. No.	Panel	Extraction thickness m	Minimum depth m	With depth ratio	Depth factor	Extraction percentage factor	Goaf treatment factor	Rock mass factor
1	2	3	4	5	6	7	8	9
1	Z	4.2	348	0.29	1.1	0.6	0.95	0.8
2	A6	4.2	322	0.31	1.1	0.6	0.95	0.8
3	A12	4.2	284	0.35	1.1	0.6	0.95	0.8
4	A16	4.2	237	0.42	1.0	0.6	0.95	0.8
5	39	4.2	195	0.51	1.0	0.6	0.95	0.8
6	33	4.2	165	0.61	1.0	0.6	0.95	0.8
7	34	4.2	128	0.78	1.0	0.6	0.95	0.8
8	42	4.2	136	0.74	1.0	0.6	0.95	0.8
9	50	4.2	97	1.03	1.0	0.6	0.95	0.6

Subsidence Related Parameters for Panels in 4 Seam

Sl. No.	Panel	Extraction thickness m	Minimum depth m	With depth ratio	Depth factor	Extraction percentage factor	Goaf treatment factor	Rock mass factor
1	2	3	4	5	6	7	8	9
1	Z	2.14	360	0.28	1.1	0.6	0.95	0.9
2	A6	2.14	334	0.30	1.1	0.6	0.95	0.9
3	A12	2.14	296	0.34	1.1	0.6	0.95	0.9
4	A16	2.14	249	0.40	1.0	0.6	0.95	0.9
5	39	2.14	207	0.48	1.0	0.6	0.95	0.9
6	33	2.14	177	0.56	1.0	0.6	0.95	0.9
7	34	2.14	141	0.71	1.0	0.6	0.95	0.9
8	42	2.14	148	0.68	1.0	0.6	0.95	0.9
9	50	2.14	109	0.92	1.0	0.6	0.95	0.9

Anticipated Subsidence Movements

The anticipated subsidence movements due to extraction of

the panels in the six seams by bord and pillar system with caving in descending order and in superimposition of the panels were as given below.

Subsidence Movements due to Extraction of Panels in 1A Seam

Sl. No.	Panel	Extraction thickness m	Minimum depth m	Width-depth ratio	Max. subsidence mm	Max. slope mm/m	Maximum strains Comp. mm/m	Maximum strains Tensile mm/m
1	2	3	4	5	6	7	8	9
1	Z	3.6	218	0.46	164	6.0	3.0	3.2
2	A6	3.6	192	0.52	205	8.0	4.0	4.2
3	A12	3.6	154	0.65	329	12.8	6.4	6.7
4	A16	3.6	107	0.93	822	26.9	13.5	14.2
5	39	3.6	65	1.54	1642	50.6	25.3	26.6
6	33	3.6	35	2.86	1642	93.9	47.0	49.4

Subsidence Movements due to Extraction of Panels in 1 Seam

Sl. No.	Panel	Extraction thickness m	Minimum depth m	Width-depth ratio	Max. subsidence mm	Max. slope mm/m	Maximum strains Comp. mm/m	Maximum strains Tensile mm/m
1	2	3	4	5	6	7	8	9
1	Z	2.75	238	0.42	254	5.3	2.7	2.8
2	A6	2.75	212	0.47	360	8.0	4.0	4.2
3	A12	2.75	174	0.57	571	14.4	7.2	7.6
4	A16	2.75	127	0.79	973	23.0	11.5	12.1
5	39	2.75	85	1.18	1410	33.2	16.6	17.4
6	33	2.75	55	1.82	1410	51.3	25.7	27.0

Subsidence Movements due to Extraction of Panels in 3B Seam

Sl. No.	Panel	Extraction thickness m	Minimum depth m	Width-depth ratio	Max. subsidence mm	Max. slope mm/m	Maximum strains Comp. mm/m	Maximum strains Tensile mm/m
1	2	3	4	5	6	7	8	9
1	Z	1.83	283	0.35	103	1.9	1.0	1.1
2	A6	1.83	257	0.39	145	2.9	1.5	1.6
3	A12	1.83	219	0.46	225	4.8	2.4	2.5
4	A16	1.83	172	0.58	404	9.6	4.8	5.0
5	39	1.83	130	0.78	639	15.2	7.6	8.0
6	33	1.83	100	1.00	845	19.0	9.5	10.0
7	34	1.83	64	1.56	835	26.0	13.0	13.7

Subsidence Movements due to Extraction of Panels in 3A Seam

Sl. No.	Panel	Extraction thickness m	Minimum depth m	Width-depth ratio	Max. subsidence mm	Max. slope mm/m	Maximum strains Comp. mm/m	Maximum strains Tensile mm/m
1	2	3	4	5	6	7	8	9
1	Z	2.93	293	0.34	366	5.9	3.0	3.2
2	A6	2.93	267	0.37	436	7.5	3.8	4.0
3	A12	2.93	229	0.44	571	10.6	5.3	5.6
4	A16	2.93	180	0.56	857	15.2	7.6	8.0
5	39	2.93	140	0.71	1127	24.1	12.1	12.7
6	33	2.93	110	0.91	1444	29.5	14.8	15.5
7	34	2.93	74	0.35	1587	42.9	21.5	22.6
8	42	2.93	81	1.23	668	20.6	10.3	10.8

Subsidence Movements due to Extraction of Panels in 3 Seam

Sl. No.	Panel	Extraction thickness m	Minimum depth m	Width-depth ratio	Max. subsidence mm	Max. slope mm/m	Maximum strains Comp. mm/m	Maximum strains Tensile mm/m
1	2	3	4	5	6	7	8	9
1	Z	2.14	348	0.29	60	1.0	0.5	0.5
2	A6	2.14	322	0.31	72	1.2	0.6	0.6
3	A12	2.14	284	0.35	121	2.2	1.1	1.2
4	A16	2.14	251	0.42	198	4.3	2.2	2.3
5	39	2.14	195	0.51	346	7.8	3.9	4.1
6	33	2.14	165	0.61	511	12.1	6.1	6.4
7	34	2.14	128	0.78	747	18.1	9.1	9.6
8	42	2.14	136	0.74	703	16.8	8.4	8.8
9	50	2.14	97	1.03	1054	32.6	16.3	17.1

Subsidence Movements due to Extraction of Panels in 4 Seam

Sl. No.	Panel	Extraction thickness m	Minimum depth m	Width-depth ratio	Max. subsidence mm	Max. slope mm/m	Maximum strains Comp. mm/m	Maximum strains Tensile mm/m
1	2	3	4	5	6	7	8	9
1	Z	2.14	360	0.28	166	2.4	1.2	1.3
2	A6	2.14	334	0.30	191	2.9	1.5	1.6
3	A12	2.14	296	0.34	268	4.3	2.2	2.3
4	A16	2.14	249	0.40	348	6.3	3.2	3.4
5	39	2.14	207	0.48	487	9.6	4.8	5.0
6	33	2.14	177	0.56	626	13.1	6.6	6.9
7	34	2.14	141	0.71	829	17.1	8.6	9.0
8	42	2.14	148	0.68	788	16.5	8.3	8.7
9	50	2.14	109	0.92	1055	21.3	10.7	11.2

Impacts of Subsidence Movements

The anticipated impacts of the subsidence movements over the panels in the six seams on the forest areas were as outlined later in this annexure. While assessing the impacts the following conditions were taken into consideration.

- The extraction in the seams will be done on panels and barrier pattern by bord and pillar system with caving.
- The extraction will be done in descending order.
- The panels in the seams will be extracted in superimposition.
- There will a time lag of about 3-5 years in the extraction of the panels in successive seams.
- The barriers between the panels shall be designed for long-term stability with a factor of safety >2.0

Since the panels in 1A Seam will be extracted in the first instance followed by other seams in descending order the impact assessment has also been done in the same order.

Impacts due to 1A Seam Extraction

The maximum anticipated subsidence movements due to extraction of the six panels in 1A Seam are as given below :

Max. Subsidence	= 164-1642 mm
Max. Slope	= 6.0-93.9 mm/m
Max. Compressive Strain	= 3.0-47.0 mm/m
Max. Tensile Strain	= 3.2-49.4 mm/m

The impacts of these movements on the forest areas will be as given below.

1. The maximum subsidence of 164-1642 mm will cause the development of depressions on the surface over the panels. Depending upon the post subsidence surface topography the rain water may accumulate in these depressions.
2. The maximum slope due to subsidences over the panels will cause the development of tilt of the order of 0.34° to 5.36°, and hence the trees in the forest areas will tilt to this extent, which is not likely to cause practically any impact on the trees.

3. The anticipated maximum tensile strains are >20.0 mm/m over panels 39 and 33 hence the width of the cracks developing on the surface over these panels will be >300 mm. For the other four panels the width of the cracks will be <300 mm. Hence, panels 39 and 33 may be extracted with stowing or any other suitable partial extraction method.

4. The development of the cracks on the surface can be expected to cause the following problems to the forest areas and the underground workings.
   - The water retaining capacity of the soils will reduce.
   - The rate of erosion of the top soil may increase.
   - The air being circulated in the depillaring areas may find way to surface leading to ineffective ventilation of the underground workings.
   - Leakage of air through goaf having some coal may enhance the process of spontaneous heating of coal, which may ultimately lead to mine fires.
   - Inundation of underground workings, if water from the surface finds way to the underground workings through the cracks.

Impacts due to 1 Seam Extraction

The maximum anticipated subsidence movements due to extraction of the six panels in 1 Seam are as given below.

Max. Subsidence = 254-1410 mm

Max. Slope = 5.3-51.3 mm/m

Max. Compressive Strain = 2.7-25.7 mm/m

Max. Tensile Strain = 2.8-27.0 mm/m

These subsidence movements will be in addition to those having already taken place due to extraction of the panels in 1A Seam. Hence, the cumulative subsidence movements after the extraction of 1 Seam will be as given below.

S.N.	Panel	Max. subsidence mm	Max. slope mm/m	Max. strains Comp. mm/m	Max. strains Tensile mm/m
1.	Z	418	11.3	5.7	6.0
2	A6	565	16.0	8.0	8.4
3	A12	900	27.2	13.6	14.3

4	A16	1795	49.9	25.0	26.3
5	39	3052	83.8	41.9	44.0
6	33	3052	145.2	72.7	76.4

The impacts of these subsidence movements on the surface in the forest areas will be as outlined below.

1. The depressions caused on the surface due to extraction of the panels in 1A Seam will further deepen due to addition of subsidence of 1 Seam. Hence, the extent of water-logging may increase. However, before commencement of depillaring in the panels in 1 Seam the subsidence related water bodies over them will be required to be drained for the safety of the underground workings.
2. The maximum tilt of the trees due to subsidence will increase to about 8.26°, which is not going to affect any tree.
3. If the cracks formed on the surface due to extraction of the panels in 1A Seam are not adequately filled at appropriate time, they will widen with the additional subsidence movements due to extraction of 1 Seam, and their width will increase beyond 300 mm over panels A16, 39 and 33. In case the cracks formed due to extraction of the panels in 1A Seam are adequately filled, the width of the cracks due to extraction in the panels in 1 Seam will exceed 300 mm over panel 33 only.
4. The impacts on the forest area and to the underground workings will further increase if suitable measures are not taken during and after the extraction of 1A Seam.

Impacts due to 3B Seam Extraction

The maximum anticipated subsidence movements due to extraction of the seven panels in 3B Seam are as given below.

Max. Subsidence	= 103-835 mm
Max. Slope	= 1.9-26.0 mm/m
Max. Compressive Strain	= 1.0-13.0 mm/m
Max. Tensile Strain	= 1.1-13.7 mm/m

These subsidence movements will be in addition to those having already taken place due to extraction of the panels in 1A and 1 Seams. Hence, the cumulative subsidence movements due to extraction of the panels in the three seams will be as given below.

S.N.	Panel	Max. subsidence mm	Max. slope mm/m	Max. strains	
				Comp. mm/m	Tensile mm/m
1	Z	521	13.2	6.7	7.1
2	A6	710	18.9	9.5	10.0
3	A12	1125	32.0	16.0	16.8
4	A16	2199	59.5	29.8	31.3
5	39	3691	99.0	49.5	52.0
6	33	3897	164.2	82.2	86.4
7	34	835	26.0	13.0	13.7

The impacts of these subsidence movements on the surface in the forest land will be as outlined below.

1. The depressions on the surface already developed due to extraction of the two upper seams will further deepen and one new depression will be developing due to extraction of panel 34. Hence, the extent and intensity of water-logging will further increase. However, before commencement of depillaring in the panels in 3B Seam all subsidence related surface water bodies will be required to be drained for the safety of the underground workings.
2. The maximum tilt of the trees in the forest areas will increase to about 9.32°, which is not likely to cause practically any problem to the trees.
3. If the cracks formed on the surface due to extraction of the panels in the overlying two seams are not adequately filled they will further widen due to extraction of the 3B Seam and this width will be >300 mm over panels A16, 39 and 33. However, if the cracks developed due to the upper seams' extraction are adequately filled at appropriate time the width of the cracks developing due to 3B Seam's extraction will not exceed 300 mm because the maximum tensile strain will be 13.7mm/m.
4. The impacts on the forest areas and the dangers to the underground workings due to the development of the cracks on the surface will further increase if suitable measures are not taken at appropriate stages of extraction in the three seams.

Impacts of 3A Seam Extraction

The maximum anticipated subsidence movements due to extraction of the eight panels in 3A Seam are as given below.

Max. Subsidence = 366-1587 mm

Max. Slope = 5.9-42.9 mm/m

Max. Compressive Strain = 3.0-21.5 mm/m

Max. Tensile Strain = 3.2-22.6 mm/m

These subsidence movements will be in addition to those having already taken place due to the extraction of the overlying 1A, 1 and 3B Seams. Hence, the cumulative subsidence movements due to the extraction of all the four seams will be as given below.

S.N.	Panel	Max. subsidence mm	Max. slope mm/m	Max. strains Comp. mm/m	Max. strains Tensile mm/m
1	Z	877	19.1	9.7	10.3
2	A6	1146	26.4	13.3	14.0
3	A12	1696	42.6	21.3	22.4
4	A16	3056	74.7	37.4	39.3
5	39	4818	123.1	61.6	64.7
6	33	5341	193.7	97.0	101.9
7	34	2422	68.9	34.5	36.3
8	42	668	20.6	10.3	10.8

The impacts of these subsidence movements on the forest area will be as given below.

1. The depressions already developed due to subsidence as a result of extraction of the three upper seams will further deepen and one new depression will develop over panel 42. Hence, the extent of water logging will further increase. However, before commencement of extraction in the panels in 3A Seam, it will be necessary to drain out all the surface water bodies from over the panels for the safety of the underground workings.

2. The maximum tilt of the trees in the forest areas will increase to 10.96°, which is not likely to cause practically any problem to the trees.

3. If the cracks formed due to extraction of the panels in the three upper seams are not adequately filled they will further widen due to extraction of 3A Seam and their width will exceed 300 mm over panels A12, A16, 39, 33 and 34. However, if the cracks developed due to the extraction of the upper seams are adequately filled at appropriate time the width of the cracks developing on the surface due to extraction of 3A Seam will exceed 300 mm over panel 34 only.

4. If appropriate mitigative measures are not taken at appropriate time the problems to the forest areas and also to the underground workings will further intensify.

Impacts of 3 Seam Extraction

Nine panels are planned for extraction in 3 Seam below forest area out of which eight are in superimposition of the panels in the upper seams and one panel, namely, 50, has intact overlying rock mass. The maximum anticipated subsidence movements due to extraction of these panels are as given below.

Max. Subsidence = 60-1054 mm

Max. Slope = 1.0-32.6 mm/m

Max. Compressive Strain = 0.5-16.3 mm/m

Max. Tensile Strain = 0.5-17.1 mm/m

These subsidence movements will be in addition to those having already taken place due to extraction of the panels in the four seams above. Hence, the cumulative subsidence movements will be as given below.

S.N.	Panel	Max. subsidence mm	Max. slope mm/m	Max. strains	
				Comp. mm/m	Tensile mm/m
1	Z	937	20.1	10.2	10.8
2	A6	1218	27.6	13.9	14.6
3	A12	1817	44.8	22.4	23.6
4	A16	3254	79.0	39.6	41.6
5	39	5164	130.9	65.5	68.8
6	33	5852	205.8	103.1	108.3
7	34	3169	87.0	43.6	45.9
8	42	1371	37.4	18.7	19.6
9	50	1054	32.6	16.3	17.1

The impacts of the above subsidence movements on the forest areas will be as given below.

1. The depressions already developed due to subsidences as a result of extraction of the four upper seams will further deepen and one new depression will develop over panel 50. Hence, the extent of water logging will increase. However, for the safety of the underground workings in the panels in 3 Seam it would be necessary to drain out all the overlying surface water bodies before commencement of depillaring.
2. The maximum tilt of the trees in the forest areas due to subsidence will increase to about 11.63°, which will not have any adverse impact on the trees.
3. If the cracks developed due to extraction of the overlying

four seams are not filled at appropriate time, they will widen further and their width may exceed 300 mm over panels A12, A16, 39, 33 and 34. However, if the cracks developed earlier are adequately filled the width of the new cracks developing due to extraction of 3 Seam will not exceed 300 mm.

4. If suitable mitigative measures are not taken at appropriate times the problems to the forest areas and also to the underground workings due to the development of the cracks will further intensify.

Impacts of 4 Seam Extraction

Nine panels are planned for extraction in this seam below the forest areas which are in superimposition of the panels in the upper seams. The maximum anticipated subsidence movements due to extraction of these panels are as given below.

Max. Subsidence = 166-1055 mm

Max. Slope = 2.4-21.3 mm/m

Max. Compressive Strain = 1.2-10.7 mm/m

Max. Tensile Strain = 1.3-11.2 mm/m

These subsidence movements will be in addition to those having already taken place due to extraction of the panels in the overlying five seams. Hence, the cumulative subsidence movements due to extraction of the panels in the six seams will be as given below.

S.N.	Panel	Max. subsidence mm	Max. slope mm/m	Max. strains	
				Comp. mm/m	Tensile mm/m
1	Z	1103	22.5	11.4	12.1
2	A6	1409	30.5	15.4	16.2
3	A12	2085	49.1	24.6	25.9
4	A16	3062	85.3	42.8	45.0
5	39	5633	140.5	70.3	73.8
6	33	6478	218.9	109.7	115.2
7	34	3998	104.1	52.2	54.9
8	42	2159	53.9	27.0	28.3
9	50	2109	53.9	27.0	28.3

The impacts of these subsidence movements on the forest areas are as outlined hereunder

1. The depressions already developed due to subsidence

movements as a result of extraction of the upper seams will further deepen. Hence, the extent of water-logging will further increase. However, for the safety of the underground workings in the panels in 4 Seam it will be necessary to drain out all the surface water bodies from over the panels before commencement of extraction in the panels. After mining has been completed in all the seams the desirable water bodies may be retained.

2. The maximum tilt of the trees in the forest land due to subsidence will increase to about 12.35°, which will not cause any problem to the trees.

3. If the cracks developed due to the extraction of the upper seams are not filled at appropriate time they will further widen and their width will exceed 300 mm over panels A12, A16, 39, 33, 34, 42 and 50. However, if these cracks are adequately filled, the width of the cracks developing due to extraction of the panels in 4 Seam will not exceed 300 mm.

4. If suitable mitigative measures are not take at appropriate times the problems to the forest areas and also to the underground workings due to the development of the cracks will further intensify.

The **problems to the underground workings** due to development of the cracks on the surface may also arise as a result of the underground mining in non-forest areas.

The **voids remaining underground** after finishing of mining in all the seams, to the extent of about 50% of the total volume extracted, will get filled with water in due course of time forming an underground water body. This can be considered as a beneficial development as the water from this water body can be used for various purposes not only in forest areas but also in non-forest areas.

IMPACT MITIGATION

For the mitigation of the impacts of the subsidence movements due to extraction of the panels in the six seams by bord and pillar system in descending order and in superimposition the following mitigative and management measures will be required to be taken.

Mitigative Measures

1. **Width of the Barrier Between the Panels**: Since the extraction in the six seams is planned by panel and barrier system the width of the barriers between the panels be designed for long-term stability with a factor of safety >2.0.

2. **Water-logging in Depressions:** There will be water-logging in the depressions developed due to subsidence movements which will result in the development of the surface water bodies. This can be considered as a welcome development as water from these water bodies can be used for various purposes in the forest areas. The decision regarding retaining of the surface water bodies should be taken when all the seams have been extracted.

3. **Protection of Underground Workings Against Surface Water Bodies:** Before commencement of depillaring in the panels it would be necessary to drain all the surface water bodies over them to ensure safety of the underground workings.

4. **Tilting of Trees in the Forest Areas:** The maximum tilt of the trees in the forest due to the extraction of the panels in the six seams will be as given below. This magnitude of the tilt due to subsidence is not expected to cause practically any problem to the trees. Hence, no mitigative measure is required to be taken on this account.

Seams	1A	(+)1	(+)3B	(+)3A	(+)3	(+)4
Max. slope (mm/m)	93.9	145.2	164.2	193.7	205.8	218.9
Max. tilt (degree)	5.36°	8.26°	9.32°	10.96°	11.63°	12.35°

(+) Cumulative maximum slope/tilt.

5. **Maximum tensile Strains:** The maximum cumulative tensile strains over the panels in the six seams will be as given below.

S.N.	Panel	Tensile strains over seams in mm/m					
		1A	(+)1	(+)3B	(+)3A	(+)3	(+)4
1	Z	3.2	6.0	7.1	10.3	10.8	12.1
2	A6	4.2	8.4	10.0	14.0	14.6	16.2
3	A12	6.7	14.3	16.8	22.4	23.6	25.9
4	A16	14.2	26.3	31.3	39.3	41.6	45.0
5	39	26.6	44.0	52.0	64.7	68.8	73.0
6	33	49.4	76.4	86.4	101.9	108.3	115.2

7	34	–	–	13.7	36.3	45.9	54.9
8	42	–	–	–	10.8	19.6	28.3
9	50	–	–	–	–	17.1	28.3

(+) Cumulative strains

It is noted that even after the extraction of all the six seams the maximum tensile strains over the panels Z and A6 will not exceed 20 mm/m. In general it can be said that for the initial planning the extraction of the following panels be taken into consideration.

Panel	Seams
Z	1A, 1 and 3B
A6	1A, 1 and 3B
A12	1A, 1 and 3B
A16	1A
34	3B
42	3A
50	3

For the other panels the following actions can be considered.

- Panels 39 and 33 in 1A Seam may be extracted with hydraulic stowing or any other suitable partial extraction method. Extraction with stowing would reduce the maximum tensile strains from 26.6 and 49.4 mm/m to 2.0 and 3.64 mm/m respectively.
- Panels A16, 39 and 33 in 1 Seam may also be extracted with hydraulic stowing or any other suitable partial extraction method. Extraction with stowing would reduce the maximum tensile strains from 26.3, 44.0 and 76.4 mm/m to 1.9, 3.2 and 5.6 mm/m respectively.
- Panels A16, 39 and 33 in 3B Seam may also be extracted with hydraulic stowing or any other suitable partial extraction method. Extraction with stowing would reduce the maximum tensile strains from 31.6, 52.0 and 86.4 mm/m to 2.3, 3.8 and 6.4 mm/m respectively.
- The decision for extraction of the other panels should be taken after the extraction of the three upper seams.

During the extraction of the panels with the above mitigative measures/restrictions cracks will develop on the surface. The following actions are being suggested for taking care of the cracks.

- The cracks should be filled with soil and stone chips and after filling a heap of soil of about 0.3 m height be made over the cracks.

- Depillaring in the panels in the lower seams should not be started till such time the cracks developed on the surface due to extraction of the corresponding panels in the upper seams have been adequately filled in the manner outlined above.
- An accurate record of the development of the cracks on the surface and their filling shall be maintained.

6. **Improving Water Retaining Capacity of Soils:** The water retaining capacity of the soils reduces due to formation of the cracks in the tensile strain zone. This can be improved by filling the cracks in the manner described earlier.

7. **Reducing Chances of Soils Erosion:** Due to the tensile strains and hence the development of the cracks on the surface the top soil loosens and hence it becomes prone to erosion. The chances for this erosion can be minimised by suitably filling the cracks.

8. **Minimising Leakage of Air from Underground Workings:** As soon as the cracks develop on the surface due to subsidence movements the underground workings get connected with the surface and in such situations the air which is being circulated for ventilation in depillaring districts may find way to the surface through the cracks. Hence, the quantity of air in the return roadways of the districts will reduce. In order to maintain adequate ventilation more air will be required for circulation in the districts. Adequate filling of the cracks will minimise leakage of air from the underground workings.

9. **Minimising Chances of Spontaneous Heating in Goaves:** Due to leakage of air through the goaves having coal left during depillaring spontaneous heating may take place which if not controlled may cause development of fires. The chances of this spontaneous heating can be minimised by minimising the leakage of air through the goaves and this can be achieved by suitably filling the cracks on the surface.

10. **Minimising Chances of Inundation:** During rainy seasons the water from the surface may find way to the underground workings through the cracks developed due to subsidence movements. This may result in an increase in the make of water in the mine and in extreme situation may cause inundation of underground workings. The

entry of water to the underground workings can be minimised by adequately filling the cracks.

11. **Underground Water Bodies:** When mining is completed in an area the voids remaining underground, to the extent of about 50% of the total volume extracted, get filled with water and thus an underground water body is formed. The water from this can be used for various purposes. Hence, this is considered to be a welcome development.

Management Measures

1. For undertaking various activities for the mitigative measures in the forest areas, e.g., survey of cracks, transport of soil and stone chips, filling of cracks, etc., the officials of the mine will be required to go and work in the forest areas. The local forest authorities shall be required to accord necessary permission for the mine officials to enter the forest areas and undertake the activities.
2. It will be advisable to the colliery management to suitably fill up the surface cracks due to subsidence for the safety of the underground workings.
3. Since in a mine having been planned for extraction with caving the formation of the cracks on the surface can be a regular continuous phenomenon it will be advisable to form a full time well equipped team for the purpose of filling the cracks and undertaking the mitigative measures.
4. The colliery management shall be required to make arrangements for adequate quantities of soil and stones chips.

RECOMMENDATIONS

On the basis of the foregoing discussions the following recommendations were made for the extraction of the panels in six Seams, namely, 1A, 1, 3B, 3A, 3 and 4, at RK7 and RK New Tech Inclines of SCCL.

1. The barrier between the panels be designed for long term stability with a factor of safety >2.0.
2. The extraction in the panels will be done by bord and pillar system in descending order and in superimposition. There will be a gap of about 3-5 years in the extraction of

panels in successive seams.

3. The following panels may be extracted with hydraulic stowing or any other suitable partial extraction method.

Seam	Panels
1A	39 and 33
1	A16, 39 and 33
3B	A16, 39 and 33

4. The following panels may be extracted with caving

Seam	Panels
1A	Z, A6, A12, A16
1	Z, 16, A12
3B	Z, A6, A12, 34
3A	42
3	42, 50

5. For the remaining panels below the forest area the decision should be taken after the extraction of the three upper seams, namely, 1A, 1 and 3B.

6. The extraction in the panels will not be started till such time the overlying surface water bodies formed due to subsidence are drained and also the cracks developed on the surface have been adequately filled in the manner outlined earlier.

7. After the mining is completed in the seams underground the desirable surface water bodies may be retained.

8. The colliery management shall be required to maintain an accurate record of the development and filling of the cracks.

9. The local forest authorities shall be required to accord necessary permission to the colliery management for undertaking the mitigative measures in the forest areas.

10. The colliery management is advised to fill up the cracks in the non-forest areas for the safety of the underground workings.

11. The colliery management is advised to form a full time well equipped team for the purpose of filling the cracks and undertaking the mitigative measures.

12. The colliery management shall be required to make arrangement for adequate quantities of soil and stone chips.

ANNEXURE - XII

Details of Naheriya Project's Case Study

At Naheriya Project of Western Coalfields Limited (WCL) it was proposed to extract 26 panels in two seams by bord and pillar system with caving below forest land. Feasibility of extraction of the panels was studied by anticipating the subsidence movements and their impacts on the forest area. The details of the study are as outlined below.

Details of The Panels

The details of the panels in Seam II and Seam IV at Naheriya Project of WCL under consideration in this study as provided by WCL were as given below.

1	Mine/Colliery	:	Naheriya
2	Company	:	Western Coalfields Limited
3	Coalfield	:	Pench and Kanhan
4	Seams	:	Seam II and Seam IV
5	Seam thickness	:	Seam II = 1.5-2.5 m Seam IV = 2.5-3.5 m
6	Extraction thickness proposed	:	Seam II = 2.5 m Seam IV = 3.0 m
7	Details of panels below forest areas	:	Given later
8	Depth of seams	:	Seam II = 63 to 173 m Seam IV = 79 to 184 m
9	Parting between seams	:	About 15-16 m
10	Method of working	:	Bord and pillar
11	Goaf support	:	Caving
12	Percentage of extraction from the panels	:	80
13	Overlying rock mass	:	Intermixed shale, clay and sandstone beds overlain by partially disturbed basalt layers of varying thickness
14	Geological disturbances	:	No major geological disturbances. However, there were some small faults
15	Gradient of the seams	:	1 in 11 to 1 in 18 (3-5°)
16	Total extractable resources	:	1,841,400 m^3 or 2,577,960 tonne

Details of the Panels Below Forest Land

SN	Panel	Seam	Depth (av.) m	Panel size mm	Extractable reserves volume m^3	Extraction thickness m	Remarks
1	SW8	II	130	150X360	108,000	2.5	Rectangular shape
2	SW7	II	140	150X330	99,000	2.5	Rectangular shape
3	SW6	II	150	270X330	81,000	2.5	Triangular
4	SW5	II	145	150X150	45,000	2.5	Square
5	SW5a	II	135	150X150	45,000	2.5	Square
6	SW5b	II	125	150X150	45,000	2.5	Square
7	SW5c	II	110	150X150	45,000	2.5	Square
8	SW4	II	148	150X150	45,000	2.5	Square
9	SW4a	II	138	150X150	45,000	2.5	Square
10	SW4b	II	128	150X150	45,000	2.5	Square
11	SW3	II	105	150X150	45,000	2.5	Square
12	SW3a	II	95	150X150	45,000	2.5	Square
13	SW2	II	100	300X480	144,000	2.5	Triangular
14	SW8	IV	145	150X360	129,600	3.0	Rectangular
15	SW7	IV	155	150X360	118,800	3.0	Rectangular
16.	SW6	IV	165	270X300	97,200	3.0	Square
17	SW5	IV	160	150X150	54,000	3.0	Square
18	SW5a	IV	150	150X150	54,000	3.0	Square
19	SW5b	IV	140	150X150	54,000	3.0	Square
20	SW5c	IV	125	150X150	54,000	3.0	Square
21	SW4	IV	163	150X150	54,000	3.0	Square
22	SW4a	IV	153	150X150	54,000	3.0	Square
23	SW4b	IV	143	150X150	54,000	3.0	Square
24	SW3	IV	120	150X150	54,000	3.0	Square
25	SW3a	IV	110	150X150	54,000	3.0	Square
26	SW2	IV	115	300X480	172,800	3.0	Triangular
			Total		1,841,400 m^3 = 2,577,960 tonne		

On the basis of the details of the panels, the subsidence related parameters of the panels taken into consideration are as given below, which were used in the anticipation of the subsidence movements due to the extraction of the individual panels. Rock mass factor for Seam II was taken as 0.5 due to the presence of Basaltic rocks in the overlying rock mass and 0.8 for Seam IV as the parting between the seams is 15-16 m and that the panels in Seam IV will be extracted after the extraction of the panels in Seam II. Depth factor was taken as 1.0 because the average depth of all the panels is <250 m. Average depth was considered because the gradient of the seams is only 3°-5°.

S.N.	Seam	Panel	Extraction thickness m	Average depth m	Width depth ratio	Rock mass factor	Extraction percentage factor	Goaf treatment factor	Depth factor
1	II	SW8	2.5	130	1.15	0.5	0.6	0.95	1.0
2	II	SW7	2.5	140	1.07	0.5	0.6	0.95	1.0
3	II	SW6	2.5	150	1.60	0.5	0.6	0.95	1.0
4	II	SW5	2.5	145	1.03	0.5	0.6	0.95	1.0
5	II	SW5a	2.5	135	1.11	0.5	0.6	0.95	1.0
6	II	SW5b	2.5	125	1.20	0.5	0.6	0.95	1.0
7	II	SW5c	2.5	110	1.36	0.5	0.6	0.95	1.0
8	II	SW4	2.5	148	1.01	0.5	0.6	0.95	1.0
9	II	SW4a	2.5	138	1.09	0.5	0.6	0.95	1.0
10	II	SW4b	2.5	128	1.17	0.5	0.6	0.95	1.0
11	II	SW3	2.5	105	1.43	0.5	0.6	0.95	1.0
12	II	SW3a	2.5	95	1.58	0.5	0.6	0.95	1.0
13	II	SW2	2.5	100	1.50	0.5	0.6	0.95	1.0
14	IV	SW8	3.0	145	1.03	0.8	0.6	0.95	1.0
15	IV	SW7	3.0	155	0.97	0.8	0.6	0.95	1.0
16	IV	SW6	3.0	165	1.45	0.8	0.6	0.95	1.0
17	IV	SW5	3.0	160	0.94	0.8	0.6	0.95	1.0
18	IV	SW5a	3.0	150	1.00	0.8	0.6	0.95	1.0
19	IV	SW5b	3.0	140	1.07	0.8	0.6	0.95	1.0
20	IV	SW5c	3.0	125	1.20	0.8	0.6	0.95	1.0
21	IV	SW4	3.0	163	0.92	0.8	0.6	0.95	1.0
22	IV	SW4a	3.0	153	0.98	0.8	0.6	0.95	1.0
23	IV	SW4b	3.0	143	1.05	0.8	0.6	0.95	1.0
24	IV	SW3	3.0	120	1.25	0.8	0.6	0.95	1.0
25	IV	SW3a	3.0	110	1.36	0.8	0.6	0.95	1.0
26	IV	SW2	3.0	115	1.30	0.8	0.6	0.95	1.0

Anticipated Subsidence Movements

The anticipated subsidence movements due to extraction of the panels in the two seams in descending order and in superimposition are as given in the tables below.

Maximum Subsidence Movements due to Extraction of the Panels in Seam II

S.n.	Panel	Extraction thickness m	Average depth* m	Width depth ratio+	Max. subsidence mm	Max. slope mm/m	Maximum strain Comp. mm/m	Maximum strain Tensile mm/m	Remarks
1	SW8	2.5	130	1.15	802	17.0	8.5	8.9	*Average depth has been taken because the gradient of the seams is only 3-5°. + With respect to average depth
2	SW7	2.5	140	1.07	716	16.6	8.3	8.7	
3	SW6	2.5	150	1.60	1069	14.3	7.2	7.6	
4	SW5	2.5	145	1.03	673	16.2	8.1	8.5	
5	SW5a	2.5	135	1.11	759	15.7	7.9	8.3	
6	SW5b	2.5	125	1.20	855	17.1	8.6	9.0	
7	SW5c	2.5	110	1.36	994	18.1	9.1	9.6	
8	SW4	2.5	148	1.01	652	15.4	7.7	8.1	
9	SW4a	2.5	138	1.09	620	12.6	6.3	6.6	
10	SW4b	2.5	128	1.17	823	16.1	8.1	8.5	
11	SW3	2.5	105	1.43	1037	19.8	9.9	10.4	
12	SW3a	2.5	95	1.58	1069	22.5	11.3	11.9	
13	Sw2	2.5	100	1.50	1069	21.4	10.7	11.2	

Maximum Subsidence Movements due to Extraction of the Panels in Seam IV

S.n.	Panel	Extraction thickness m	Average depth* m	Width depth ratio+	Max. subsidence mm	Max. slope mm/m	Maximum strain Comp. mm/m	Maximum strain Tensile mm/m	Remarks
1	SW8	3.0	145	1.03	1409	21.9	11.0	11.6	*Average depth has been taken because the gradient of the seams is only 3-5°. + With respect to average depth.
2	SW7	3.0	155	0.97	1340	21.6	10.8	11.3	
3	SW6	3.0	165	1.45	1540	18.7	9.4	9.9	
4	SW5	3.0	150	0.94	1294	20.2	10.1	10.6	
5	SW5a	3.0	150	1.00	1386	20.8	10.4	10.9	
6	SW5b	3.0	140	1.07	1432	21.8	10.9	11.4	
7	SW5c	3.0	125	1.20	1540	24.6	12.3	12.9	
8	SW4	3.0	163	0.92	1263	19.4	9.7	10.2	
9	SW4a	3.0	153	0.98	1355	19.9	10.0	10.5	
10	SW4b	3.0	143	1.05	1425	20.9	10.5	11.0	
11	SW3	3.0	120	1.25	1540	25.7	12.9	13.5	
12	SW3a	3.0	110	1.36	1540	28.0	14.0	14.7	
13	Sw2	3.0	115	1.30	1540	26.8	13.4	14.1	

Impacts of Subsidence Movements

It is noted from the details of the panels, given earlier, that their extraction in superimposition will be done in descending order and there will a time gap of about 3-5 years in their extraction. Keeping this in view the impacts on the forest area due to the subsidence movements will be as outlined hereunder.

1. The surface area in the forest land will subside twice at an interval of about 3-5 years. The first time the subsidence will be due to extraction of Seam II and the second time due to extraction of Seam IV. At the first time the impacts will be due to the subsidence movements due to extraction of Seam II and the second time the impacts will be due to the cumulative movements as a result of extraction of both the seams.

2. **Impacts of Extraction of Seam II**: The maximum anticipated subsidence movements due to extraction of the panels in Seam II are as given below:

Maximum subsidence	= 620-1069 mm
Maximum slope	= 12.6-22.5 mm/m
Maximum comp. strain	= 6.3-11.3 mm/m
Maximum tensile strain	= 6.6-11.9 mm/m

The impacts of these subsidence movements on the forest land will be as outlined hereunder.

(a) The subsidence movements over the panels in Seam II will cause development of depressions on the surface and depending upon the original topography and the topography resulting after subsidence the rain water may accumulate in these depressions.

(b) The maximum tilt of the ground and hence of the trees in the forest areas due to a maximum slope of 22.5 mm/m will be about 1.29°, which will not cause practically any impact on the trees.

(c) The maximum tensile strains varying between 6.6 and 11.9 mm/m will cause development of cracks of up to about 150-160 mm width, i.e., the width of the cracks on the surface will not exceed 300 mm.

(d) The development of the cracks on the surface can be expected to cause the following problems to the forest area and the underground workings.

(i) Problems to the forest areas

- reduction in water holding capacity of the top-soil and sub-soils, and
- increased erosion of the top soil.

(ii) Problems to the underground workings

- leakage of ventilating air from the panels under depillaring,
- spontaneous heating/fires in the coal left in caved goaves, and
- inundation of underground workings.

3. **Impacts of Extraction of Seam IV**: The maximum anticipated subsidence movements due to extraction of Seam IV panels are as given below.

Maximum subsidence	= 1263-1540 mm
Maximum slope	= 18.7-28.0 mm/m
Maximum comp. Strain	= 9.4-14.0 mm/m
Maximum tensile strain	= 9.9-14.7 mm/m

These subsidence movements will be in addition to the subsidence movements already taken place due to extraction of the panels in Seam II. Hence, the cumulative subsidence movements after the extraction of both the seams will be as given below.

Panel	Maximum subsidence mm	Maximum slope mm/m	Maximum strain Comp. mm/m	Maximum strain Tensile mm/m
SW8	2211	38.9	19.5	20.5
SW7	2056	38.2	19.1	20.0
SW6	2609	32.9	16.6	17.5
SW5	1967	36.4	18.2	19.1
SW5a	2145	36.5	18.3	19.2
SW5b	2287	38.9	19.5	20.4
SW5c	2534	42.7	21.4	22.5
SW4	1915	34.8	17.4	18.3
SW4a	1975	32.5	16.3	17.1
SW4b	2248	37.0	18.6	19.5
SW3	2577	45.5	22.8	23.9
SW3a	2609	50.5	25.3	26.6
SW2	2609	48.2	24.1	25.3

The impacts of these subsidence movements will be as outlined hereunder.

(a) The depressions on the surface developed as a result of the extraction of Seam II will further deepen and hence the magnitude of rain water accumulation in these depressions will increase.

(b) The maximum tilt due to subsidence will increase to about 2.89°, which is also not expected to cause practically any impact on the trees in the forest area.

(c) If the cracks formed due to extraction the panels in Seam II are not adequately filled, their width will increase due to extraction of Seam IV and will exceed 300 mm over panels SW8, SW7, SW5b, SW5c, SW3, SW3a and SW2. However, if the cracks due to Seam II extraction are suitably filled the width of the cracks due to Seam IV extraction will not exceed 300 mm as the maximum anticipated tensile strains due to this extraction are <20 mm/m.

(d) The intensity of the other impacts, i.e., to the forest area and the underground workings as outlined earlier will increase if suitable precautionary and mitigative measures are not taken during and after the extraction of Seam II.

4. The underground extraction of the coal seams leaves voids of about 50% magnitude of the total volume extracted. In time these voids will become waterlogged to form an underground water body. In the long run such water bodies will be useful not only in forest areas but also in other areas.

IMPACT MITIGATION

For taking care of the impacts of the subsidence movements due to extraction of the two seams the following mitigative and management measures will be required to be taken. While suggesting these measures it is being assumed that the extraction of the two seams in descending order will be done by panel and barrier system and in superimposition.

Mitigative Measures

1. The width of the barrier between the panels in both the seams be designed for long-term stability with a factor of safety >2.0.
2. The cracks developing due to extraction of Seam II should be suitably filled with soil and stone chips and then making about 0.3 m high heap of soil over the filled cracks. This will help in

- improving water retaining capacity of the top soil,
- minimising erosion of the soils,
- improving ventilation in depillaring panels,
- minimising chances of spontaneous heating of coal left in the goaves, and
- avoiding chances of inundation of underground workings.

3. It will be necessary to fill up the cracks again during extraction of Seam IV.
4. No mitigative measure is required to be taken against tilting of the trees in the forest land as the maximum tilt will be of the order of 2.89°.
5. Waterlogging in the depressions caused due to subsidence movements is a beneficial impact and therefore no mitigative measure is required in this respect also. However, during extraction of the panels in Seam IV it will be necessary to drain out all the water bodies over them for the safety of the underground workings. In the end after the extraction of all the seams the desired surface water bodies developed due to subsidence may be retained.
6. The underground water body developed due to mining is a beneficial development. Hence, no mitigative measure is required on this account also.
7. The mine management will be required to maintain an accurate record of the development and filling of the cracks.

Management Measures

1. For implementing the mitigative measures the colliery management will be required to take various actions in the forest land, e.g., inspection and monitoring, transport of soil and stone chips, etc. The local forest authorities shall be required to accord necessary permission to the mine management for undertaking the above actions in the forest areas.
2. Mining with caving in the non-forest areas will also cause development of cracks on the surface. For the safety of the underground workings the colliery management will be required to fill up these cracks also.

3. It is being suggested that the colliery management may form an exclusive well equipped team for filling the cracks.
4. The colliery management will be required to arrange for the required quantities of soil and stone chips.

RECOMMENDATIONS

The following recommendations were made for the extraction of the panels in Seam II and Seam IV by bord and pillar system with caving in superimposition and in descending order with a time lag of 3-5 years.

1. In the seams, the barriers in superimposition, should be designed such that they have long-term stability with a factor of safety > 2.0.
2. The cracks developed due to extraction of successive seams will be filled as they develop with soil and stone chips and then a heap of about 0.3 m height be made over the filled cracks.
3. It will be necessary to fill the cracks after extraction of each seam and in case this is not done the extraction in the lower seam shall not be started, as this is not only required for surface protection but is also essential for the safety of the underground workings.
4. For the safety of the underground workings it will be necessary to drain all the subsidence related water bodies before commencement of extraction in Seam IV. However, desirable water bodies may be retained after the extraction of all the seams.
5. The situation in respect of the development of subsidence movements and their impacts and the effectiveness of the mitigative measures, outlined earlier, should be reviewed periodically. If necessary, suitable corrective measures be taken.
6. The local forest authorities shall be required to accord necessary permission to the mine management for undertaking the actions for implementing the mitigative measures in the forest areas.
7. The colliery management shall take management measures as suggested in this report.

CHAPTER 3

SUBSIDENCE MANAGEMENT IN THE AREAS HAVING GROUND WATER WITHDRAWAL WITH SPECIAL REFERENCE TO INDIAN SITUATIONS

A. PREAMBLE

It is an established fact that ground water withdrawal for domestic and industrial purposes causes subsidence movements and many areas in the world have experienced these movements and their impacts. The main problems likely to be caused due to these movements are —

- damages to the buildings and structures,
- flooding of areas,
- development of sink-holes, etc.

Although ground water withdrawal in India has been going on since ages, the subsidence aspect of this activity has not been addressed adequately. The areas likely to experience this subsidence, as identified by the author, are outlined in Table 1.11, which indicates that the entire Indo-Gangetic plane and many cities are either already experiencing this subsidence or may experience it in the near future. It is therefore, advisable that suitable action for the management of such subsidence is taken urgently.

B. PHENOMENON

It is generally believed that land subsidence arising from

ground water withdrawal is essentially the response of the surface to the compaction of the sediments holding water (aquifers) at depths. Withdrawal of water reduces the water pressure in the pores, which in turn increases effective vertical stress and thus causes compaction of the solid structure of the aquifers. This results in an increase in effective stress (grain to grain load) in the aquifer system causing large volumes of land subsidence (Prudic and Williamson, 1984). The following characteristics of water are responsible for the land subsidence phenomenon due to ground water withdrawal.

- Water cannot be compressed and has different properties of strain under the action of forces. When compressed water moves-up in the layers of the aquifers and it is expanded as it goes down.
- Water has properties of flowing and transmitting static hydraulic stresses.

Hence, the changes in ground water level are related to the quantum of ground water withdrawal and are a function of stresses in the aquifers.

Generally speaking water in an undisturbed aquifer remains at a certain hydrostatic pressure as shown in Fig. 3.1. When ground water withdrawal is started through a bore hole the hydrostatic pressure around the bore hole reduces due to suction action by the pump. This causes a drawdown in hydrostatic pressure profile of the aquifer as seen in Fig. 3.2. As a result of the changes in the pressure profile water in the aquifer moves from high pressure zones to low pressure area resulting in lowering of

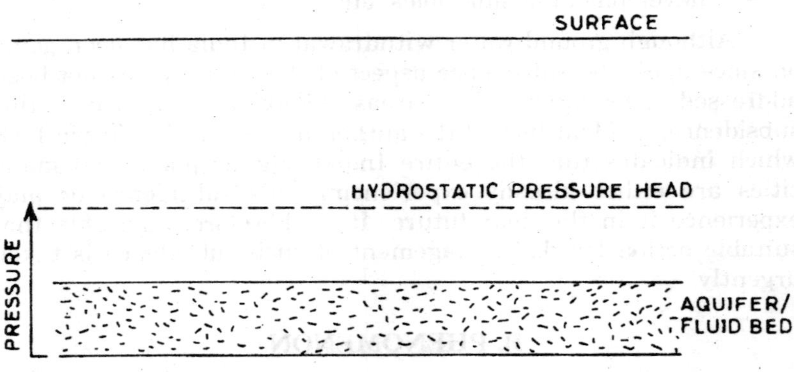

Fig.3.1. Hydrostatic pressure in an undisturbed aquifer

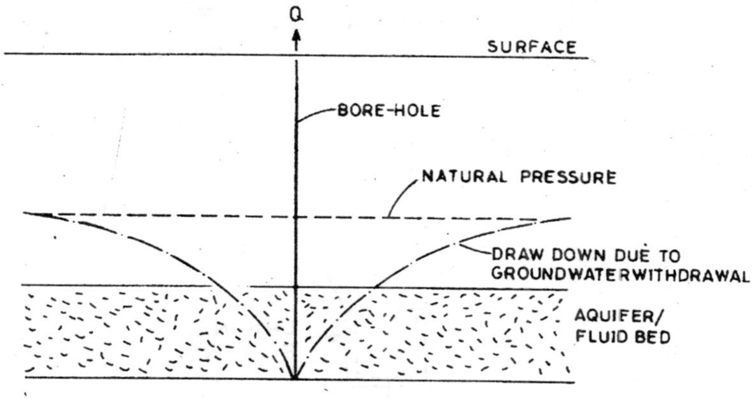

Fig. 3.2. Draw down in hydrostatic pressure due to pumping from one bore-hole

the water level in the aquifer. In areas having ground water withdrawal from a large number of bore holes, a large zone of low pressure is created (Fig. 3.3). This results in a situation in which in the depleted zones of the aquifers the vertical stress due to depth which is invariably more than the lateral stress starts compaction of the aquifers and with continued ground water withdrawal the compaction continues. Initially, as is usual with most rocks, the compaction takes place under elastic condition but with excess withdrawal the elastic compaction state is followed by permanent deformation, when the rate of compaction tends to increase. The elastic compaction process is reversible and beyond this the process is irreversible

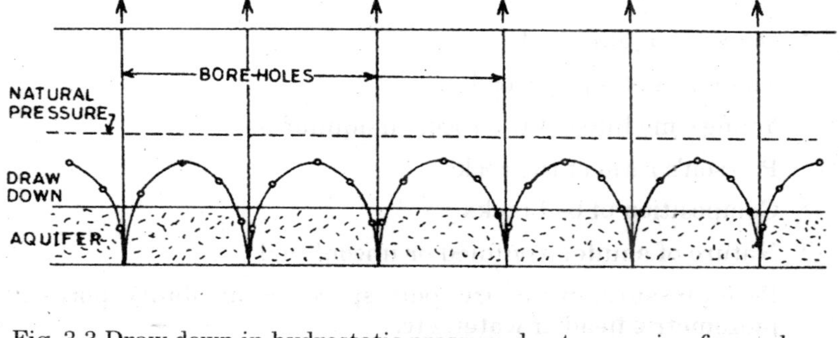

Fig. 3.3.Draw down in hydrostatic pressure due to pumping from a large number of bore-holes

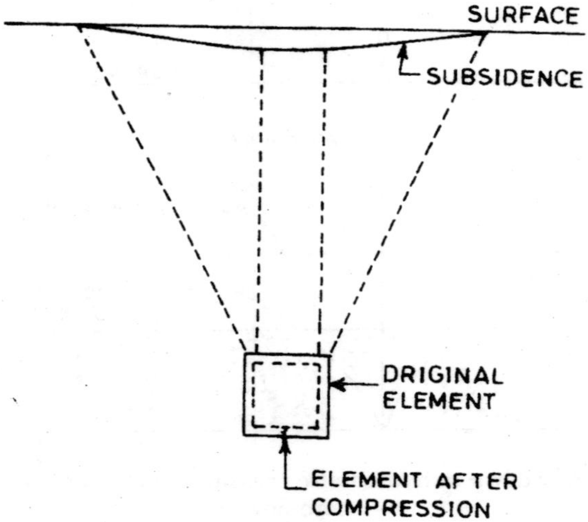

Fig. 3.4. Compaction of an element and subsidence

The aquifer compaction is aided by a unique property of water that it dissolves the soluble solids available in the aquifer rocks, which tends to increase pore space. Also, when water is pumped out, it carries with it very fine solid particles in suspension, which also tends to increase the pore space.

The compaction of the aquifer is reflected on the surface in the form of subsidence movements (Fig. 3.4) and depending upon the depth of aquifer and the characteristics of the superincumbent rock mass the transfer of the aquifer compaction to surface varies. The following factors are responsible for land subsidence due to ground water withdrawal.

Natural Factors

* Depth of aquifer bed
* Thickness of aquifer bed
* Young's modulus of the rocks in aquifer bed.
* Poisson's ratio of the rocks
* Composition of bed rocks
* Nature of aquifer, confined or unconfined
* Pore pressure, grain size, pore space, permeability, porosity, piezometric head of water, etc.

Operational Factors

* Location of bore-hole(s)
* Rate of ground water withdrawal
* Changes in pore pressure
* Lowering of piezometric head
* Rate of recharging, etc.

It has generally been found that the magnitude of surface subsidence is proportional to the drawdown of piezometric level in the aquifers.

Different types of rocks respond differently to the changes in the surrounding stress field. The magnitude of compaction of the aquifer beds and hence the land subsidence depends on the compaction characteristics. In addition to the aquifer beds, compaction of the overlying/underlying beds as a response to the changes in the stress field contributes to the subsidence.

Land subsidence in general is not influenced by the composition of the superincumbent strata. It is influenced to some extent by the position of the rigid basement below the reservoir. This subsidence has been found to be directly proportional to the rate of ground water withdrawal (Sato, et al., 1984 and Daito, et al., 1991).

C. IMPACTS

In the Indian context the impacts of the land subsidence due to ground water withdrawal could be as outlined hereunder.

Impacts in Cities- Land subsidence in the cities, e.g., Kolkata, Delhi, etc., can cause the following problems.

- Changes in surface topography resulting in alterations in the surface drainage pattern.
- Damages to buildings and structures.
- Water-logging in the depressions caused by the subsidence.
- Flooding from nearby water bodies due to ground level lowering to the extent of reaching below the HFLs.

Impacts in Agricultural Areas of Indo-Gangetic Plane - The area in the country extending from Jammu in the north to West Bengal in the east has a number of rivers and is one of the most productive agricultural area. The plane is having a large number of tube-wells for meeting the water requirement in the agriculture sector. Land subsidences due to ground water withdrawal can cause the following problems in these areas.

- Depressions around the tube-wells altering the surface drainage pattern.
- Water logging in the depressions.
- Flooding from the nearby water bodies.
- Flood water may not recede from the depressions causing problems in agricultural activities.

D. LAND SUBSIDENCE MANAGEMENT

For the management of the land subsidence due to ground water withdrawal the following strategy is suggested, with special reference to the Indian situation.

1. Collection and compilation of the details

The following details of the area under study should be collected and compiled.

Hydrogeological information- This would require comprehensive study of the geology and the hydrological regime in the areas under consideration. The study is expected to reveal the following.

* General geological sequence of strata with locations of the aquifers (confined and unconfined).
* Aquifer details, including spread, thickness, rock type, porosity, permeability, connectivity to surface water source, piezometric head, aquifer water head, quality of water, quantity of water available, etc.
* Macro and micro geological structure of the area.
* Surface topography and drainage pattern.
* Present land-use pattern.
* Rainfall pattern.
* Flooding details from past case studies, i.e., history of flooding.

Activities taking place on the surface, their water requirements and the sources from which these requirements are being met. The sources may be from surface water bodies and also underground water bodies.

Water balance of the area involving a study of the water availability and its requirement to assess the adequacy of water availability.

Details of water being pumped out from aquifers are to be compiled to indicate the following.

* Locations of the tube-wells
* Aquifers from which the tube-wells are drawing water.
* Quantity of water being pumped out and its rate.
* Quality of water and its use. In these details all the tube-wells are to be covered.

2. Prediction of land subsidence

With the details outlined above the subsidence movements due to ground water withdrawal from each tube-well should be predicted. For this purpose many models are available in the literature and the most suitable one can be chosen. The following parameters are required to be predicted.

1. Maximum subsidence
2. Rate of subsidence
3. Spread of subsidence
4. Subsidence contours
5. Changes in surface contours and post subsidence contours

The prediction of the above parameters is required to be done in a time-frame at intervals of five years. For the prediction of the changes in the surface contours and hence the post subsidence contours it is essential that precise existing surface contours at least at 1.0 m, if not 0.5 m, intervals are available because generally the subsidence movements due to ground water withdrawal take place at a rate of a few centimeters per year.

The prediction should be done for a period ranging from about 25 years to 50 years.

3. Monitoring

As is well known ground water withdrawal in the locations under review is going on for years and there is a possibility of some land subsidence having already taken place or is taking place. The study of the details outlined earlier can indicate such areas. For these areas and also for those having future land subsidence potential it would be advisable to plan for monitoring of land subsidence movements.

Monitoring of the subsidence movements can be done by using any of the following well established systems.

1. Precise leveling using parallel plate micrometer geodetic and self aligning levels.
2. EDM traversing with precise leveling.
3. "Total Station" system
4. Global positioning system (GPS)

At many places in the world land subsidences are being monitored using various systems. The objective is to achieve the best possible accuracy. The selection of the systems should be done on the merits of their usefulness in the given situations.

4. Impact Assessment

After the land subsidence movements have been predicted it would be necessary to assess the impacts that these subsidence movements are likely to have on the surface at various stages. Hence, the predicted subsidence data is studied for the following.

1. Assessment of the changes in the surface drainage pattern.
2. Assessment of the chances of water logging in the depressions caused due to subsidences.
3. Assessment of the flood proneness of the subsided areas.
4. Damages to the buildings and structures.
5. Impacts on canals, main drains and rivers.

The assessment of the changes in the surface drainage pattern, the chances of water-logging and flood proneness is done by the study of the post subsidence surface contours. Since these are rather sensitive and important issues, accurate prediction is necessary. The post subsidence surface contours should for this purpose be predicted at 0.5 to 1.0 m intervals.

The assessment of damages to the buildings and structures can be done by studying the pattern of the subsidence movements to which these will be subjected. Depending upon the nature and magnitude of the subsidences and the nature of the constructions the damages may vary.

For studying the impacts on canals, drains and rivers it is required to predict changes in their profiles due to subsidence which may cause over flow of water from the sides/embankments and flooding of nearby areas.

In the locations near sea coasts and rivers having tidal phenomenon the study should include the possibilities of flood proneness during high tides.

To the best of the knowledge of the author such studies have not been reported for the Indian conditions.

5. Management measures

The management measures in respect of the subsidences due to ground water withdrawal are as outlined hereunder.

1. Controlled withdrawal of water from the underground sources, i.e., pumping of only the bare minimum required quantity of water from aquifers and water table. This would also involve measures to avoid spillage of water.
2. Recycling the water on the surface by adopting suitable water conservation and treatment strategies. In cases where excess water is available after recycling this can be charged back into the aquifers.
3. Developing surface water resources and augmentation of water availability from the existing surface water sources.
4. Developing underground water bodies, specially in the areas having underground mining as has been discussed in the previous chapter.
5. Registration of all the tube wells to the aquifers and maintaining records of ground water withdrawal.
6. Prediction of land subsidence due to ground water withdrawal and its impacts on the surface topography. If the impacts indicate formation of depressions, then making arrangements for the drainage of water from such areas.
7. Monitoring of the subsidence movements to assess the applicability of the prediction methodologies and development of the most suitable methodologies for the Indian situations.

6. Concluding Remarks

In view of the above discussions it is being opined that the areas having ground water withdrawal in India require urgent attention from the stand point of subsidence management. Extended delays may cause serious problems in future.

BIBLIOGRAPHY

AIMM (1973): Subsidence in Mines.Proc. Symp. of Aust. Inst. Min. & Met., Wollonyong.

AIT (1982): Investigation of Land Subsidence Caused by Deep Well Pumping in the Bangkok Area. Phase IV. AIT Research Report No.161, Bangkok, pp 1 - 98.

Abel, L. and Sun, M. (1968): Ground Control in Bedded Foundations. Bull.28 Research Division, Virginia Polytechnic Institute.

Abel, J.F. and Gentry, D.W. (1978): A longwall subsidence prediction model. Proc. ASCE Spring Convention Coal Mine Subsidence, Pittsburgh, Pennsylvania, pp 56 - 76.

Abel, J. F. and Lee, F. T. (1983): Lithological controls on subsidence. SME of AIME Trans., v 274, pp 2028 - 2034.

Abou - Sayed, A.S. (1982): Elastic-plastic compaction model applicable to low cohesion oil sands reservoirs. Forum - Subsidence Due to Fluid Withdrawal Fountain Head SR, Oklahoma, USA.

Adamek, V. and Jeran, P. W. (1981): Evaluation of existing predictive methods for mine subsidence in the US. Proc. 1st. Annual Conf. Gr. Cont. in Mining, Morgantown: West Verginia Univ., pp 209 - 219.

Adamek, V. and Jeran, P. W. (1982): Evaluation of surface deformation characteristics over longwall panels in the Northern Appalachian Coalfield. In State of art of Gr. Cont. in Longwall Mining and Mining Subsidence, Conf. of Society of Mining Engrs., NewYork, 16 p.

Adamek, V. and Jeran, P. W. (1985): Pre-calculation of Surface Deformations in the North Applachian Coal Region. US Bureau of Mines, IC 9042, Mine Subsidence Control, pp 34-56.

Adamek, V., Jeran, P. W. and Trevis, M. A. (1987): Prediction of Surface Deformations Over Longwall Panels in the North Applachian Coalfields.USBM, RI 9142, 19 p.

Adamek, V., Jeran, P.W., and Trevis, M.A. (1988): Perdiction of mining induced subsidence over longwall panels in the nothern Appalachian coal region, USA. Proc. CONFERENCE 1988, Fredericton, New Burnswick, Canada, pp 7 - 16.

Agioutantis, Z. and Karmis, M.(1986): A study of roof caving in the eastern U.S. coalfields. Proc. 5th Conf. on Gr. Cont. in Mining, pp 1 - 142.

Akimov, A.G.(1958): On methods of precalculating ground surface movements. Ugol., n 2, pp 20 - 23.

Aley, T.J., Williams, J.H. and Massello, S.W. (1972): Ground water contamination sinkhole collapse induced by leaky impoundments in soluble rock terren. Missouri Geological Survey and Reources, Engineering Geology Series, n 5, 32 p.

Allen, A. S.(1978): Basic questions concerning coal mine subsidence in the United States. Asso. Engg. Geologists Bull., v 15, n 2, pp 147-161.

Allgaier, F.K.(1981): Subsidence monitoring over western coal mines. Proc. 1st Workshop on Surface Subsidence due to Underground Mining, Margantown, WV,USA, pp 156 - 163.

Allgaier, F.K. (1982): Surface subsidence over longwall panels in the western United States. Proc. State of Art of Gr. Cont. in Mining and Mining Subsidence, pp 199 - 209.

Amin, A. and Bankher, K. (1996) : Review on land subsidence in Saudi-Arabia, Land Subsidence, IAHS Publication No.234, pp 19-26.

Anderson, O. B. and Remmer, O (1982): Non-random effects in the Finnish levelling of high precision. Manuscripta Geodatica, v. 7, n 4, pp 353 - 373.

Andronopoulos, B., Rozos, D. and Hadzinakos, I. (1991): Subsidence phenomenon in the industrial area of Thessaloniki, Greece. Land Subsidence, IAHS Publication No 200, pp 59 - 70

Andros, S. O. (1914a): Coal mining practice in District 1 (longwall). Mining Investigation Bull. 5. Champaign, Illinois State Geological Survey.

Andros, S. O. (1914b): Coal mining practice in District VI (mines in bed 6 in Franklin, Jacson, Perry, and Williamson Counties). Mining Investigation Bull. 8. Champaign, Illinois State Geological Survey.

Anikeev, A.V. and Fomenko, I.K. (1995) : Subsidence - sinkhole development in sand deposits above karst masses. Land Subsidence, IAHS Publication No.234, pp 27-34.

Anon (1957): Observations of surface movements under influence of mining and their theoritical interpretation. Proc. Europ. Cong. Gr. Movement, Leeds, pp 210 - 218.

Anon (1959): Subsidence Characteristics of Low Density Salty Soil in Areas of Upper Meeker Canal-Missouri River Basin Project. United States Bureau of Reclamation, Laboratory Report EM-545. Denver, Colorado, pp 1 - 7.

Anon (1965): Surface Ground Movements Associated With the Total and Partial Extraction of Stratified Mineral Deposits. M.S. Thesis, Univeristy of Nottingham, 167 p.

Anon (1969): Land Subsidence. Proceeding of the 1st International Symposium on Land Subsidence, Tokyo, Japan, 611 p.

Anon (1975): Subsidence Engineer's Handbook. National Coal Board, Mining Department, U. K., 111 p.

Anon (1977): Land Subsidence. Proceedings of the 2nd International Symposium on Land Subsidence, Anaheim,CA. USA, 670 p.

Anon (1977-78): Coal Mining Under Stored Water. Report of the Commissioner on an Enquiry into coal Mining under or in the Vicinity of stored waters of the Nepean, Australia.

Anon (1979): Report of the Committee on Safety in Coal Mines. Department of Coal, Government of India, 209 p.

Anon (1984): Status of Mine Subsidence Research in India. CMRS, Dhanbad, India, 30 p.

Anon (1986): Remedial Measures for Cracked Buildings in Expansive Soil Area. CBRI, Building Research Note 22, pp 1 - 2.

Anon (1986): Report of the Committee on Subsidence in Collieries in West Bengal, India, 166 p.

Anon (1987): Optimisation of Design of Mine Pillar Parameters and Feasibility of Extraction of Locked-up Coal Below Built-up Structures. CMRS, Dhanbad. India, 100 p.

Anon (1991): Surface Subsidence in Mining Areas. Project Report (1964 -1990), Central Mining Research Station, Dhanbad, India.

Anon (1964): On certain linear and nonlinear strata theoretical models. Proc.Int.Conf.Strata Control, New York, pp 384-394.

Arcamone, J., Schrocter, P. and Dejean, M.J.P. (1986): State of the art of mining subsidence in France. Proc. 88th Annual Gn. Meeting of C.I.M. Montreal, pp 1 - 17.

Archibald, G. I. and Weir, A. M. (1969): A field study of subsidence. Chartered Surveyor, pp 177 - 186.

Arkin, I. (1984): A sinkhole near Damascus gate, old city of Jerusalem. Land Subsidence, IAHS publication No 151, pp 565 - 574.

Asian Institute of Technology (1979): Investigation of Land Subsidence Caused by Deep Well Pumping in the Bangkok Area, Phase I, Final Report, AIT Research Report No.87, Bangkok, 225 p.

Asian Institute of Technology (1980): Investigation of Land Subsidence Caused by Deep Well Pumping in the Bangkok Area, Phase II, Final Report AIT Research Report No.89, Bangkok, 196 p.

Asian Institute of Technology (1981): Investigation of Land Subsidence Caused by Deep Well Pumping in the Bangkok Area, Comprehensive report 1978-1981.AIT Research Report No.91, Bangkok, 365 p.

Aston, T. R. C., Tammemagi, H. Y. and Poon, A. W. (1977): A review and evaluation of empirical and analytical subsidence prediction techniques. Mining Sci. & Tech., v 5, pp 55 - 69.

Astin, J. (1968): A visco-elastic analysis of ground movement due to an advancing coal face. J. Engg. Math., pp 9 - 22.

Astori, B. and Bezoari, G.(1989): Experiences of land subsidence measurements in Italy and related problems. Land Subsidence, Oxford & I B H Pub., New Delhi, pp 259 - 270.

Atkins, A.S. et.al. (1986): Integrated methods for colliery waste disposal and ground control and safety. Proc. Symp. Gr. Movement and Control Related to Coal Mining, Parkville, Australia, pp 253 - 262.

Attewell, P.B. and Farma, I.W. (1974): Ground Disturbances Caused by Shield Tunneling and Their Effects on Pipelines and Structures. Blackie, Glasgow.

Attewell, P.B. and Farmer, I.W. (1974): Ground disturbance caused by shield tunneling in a stiff cover consolidated clay. Engg. Geol., pp 361 - 381.

Augath, W. and Strerath, M. (1995): Actual geodetic monitoring of land subsidence - the changing concept of the German North Sea Coast Levelling Nẹt. Land Subsidence, IAHS Publication No 234, pp 95-102.

Aynsley, W.J. and Hewitt, G. (1961): Subsidence observations over shallow workings, including pneumatic stowing and rapidly advancing faces. Min. Engr., pp 552 - 569.

Backwith, G. H., Slemmons, D. B. and Weeks, R. E. (1991):Use of low sun angle photography for identification of subsidence induced earth fissures. Land Subsidence, IAHS Publication No 200, pp 261 - 270.

Bahuguna, P.P. (1993): Development of Mines Subsidence Prediction Model for Indian Coalfields. Ph.D. Thesis, University of Rookee, Rookee, 177p.

Bahuguna, P.P., Singh, B., Srivastava, A.M.C. and Saxena, N. C. (1991a): An empirical method for calculation of maximum subsidence. Rock Mechanics as a Multidesciplinary Science, pp 801 - 809.

Bahuguna, P.P., Singh, B., Srivastava, A.M.C. and Saxena, N. C. (1991b): A critical review of mines subsidence prediction methods. Mining Sci. & Technology, pp 369 - 382.

Bahuguna, P. P., Singh, B., Srivastava, A.M.C. and Saxena, N. C. (1991c): MULSIM/BM and MSEAMS - two computor programs for structural analysis of underground mines excavations. Proc.of the Workshop on Numerical Methods in Rock Excavation Design, NIRM, Kolar, pp I-17 - I-25.

Bai, M., Elsworth, D. and Saperstein, L. W. (1989): Prediction of surface movements with emphasis on horizontal deformation due to mining, Proceedings of the 30th US Symposium on Rock Mechanics, pp 731 - 738.

Balestri, N. and Villani, B. (1991): Study of land subsidence in the Bolognese area. Land Subsidence, IAHS Publication No 200, pp 71 - 80.

Baligh, M. M. and Fuleihan, N. F. (1978): Consolidation theory with stress reduction due to settlement.J.Geotechnical Eng. ASCE.

Banerjee, J. P., Samanta, S., Saxena, N. C. and Singh, B. (1989): Extraction below water bodies at Surakachhar. Land Subsidence, Oxford and IBH Pub., New Delhi, pp 431 - 440.

Banerjee, S. P., Sinha, S. N. and Prasad, A. S. (1986): Status of land use and restoration of mined lands in Jharia coalfield. Proceedings of the Colloquium on Water pollution and Land Reclamation in Mining Areas, Dhanbad, India, pp 21 - 27.

Bansekom, W. A. van (1986): Measurement above the Groningen gasfield (Dutch). J. II. Archit. and Civil Techn.

Baranova, V. (1996) : Recent geodynamics of the North China Basin. Land Subsidence, IAHS Publication No 234, pp.35-40.

Barbalata, J. C. (1988): An analytical projective method for deformation measurements. Proc. CONFERENCE 1988, Fredericton, New Burnswick, Canada, pp 39 - 49.

Barbarella, M., Bitelli,G., et.al. (1988): Deformation surveys of landslide using terrestrial measurement and space technique.Proc. 5th. Int. Symp. on Deformation Measurements, Vencover.

Barbarella, M., Bitelli, G., Gubellini, A., Russo, P., et.al. (1988): Deformation survey of landslides using terrestrial measurements and space techniques. Proc. CONFERENCE 1988, Fredericton, New Burnswick, Canada, pp 50 - 61.

Barends, F.B.J. (1985): Costal oriental subsidence prediction, Main Report, The Hague, pp 17 - 18.

Barends, F. B. J.(1989): Man-induced indirect land- subsidence of alluvial plains. Land Subsidence, Oxford & IBH Pub., New Delhi, pp 91-102.

Barends, F.B.J., Poel, J.T.van der and Teunissen, J.A.M. (1995) : Geomechanical simulation of static and dynamic subsidence by reservoir compaction. Land Subsidence, IAHS Publication, No.234, pp 237-246.

Bauer, R. A. and Hunt, S. R. (1981): Profile, strain and time charactaristics of subsidence from coal mining in Illinois. Proc. Workshop on Sur. Sub. due to Underground Mining, Morgantown, WV, pp 207 - 219.

Bausekom, W. A. Von(1986): Measurement above the Groningen gasfield (Dutch).J. Int. Archit. and Civil Techn. , n 6.

Bawden, W. F. and Mottahed, P. (1986): Comparison of the Subsidence Prediction Techniques Applied to Saskatchewan Potash Mining. 88th Annual Gn. Meeting of C.I.M.,Monterial, p 34.

Beck, B. F. (1984): Sinkhole development in South Georgia and Florida,U.S.A. and the founding of the Florida Sinkhole Reasearch Institute, Land Subsidence, IAHS Publication No 151, pp 607 - 616.

Beevers, C. and Wondell, K. (1954-55): Recent research in mining subsidence. Trans. I. Min.E., pp 223 - 244.

Bell, J.W. (1991): Elevation changes associated with subsidence in Las Vegas Valley, Nevada. Land Subsidence, IAHS Publication No 200, pp 473 - 482.

Bekendom, R.G. and Vliet, A.J.J.van (1995) : Delineation of solution cavities and subsidence evaluation by means of high resolution 3D seismic reflection in Twente, East Netherlands. Land Subsidence, IAHS Publication, No.234, pp.247-258.

Benito, G. and Campo, P. Perez Del. (1991): Sinkhole evolution in alluvial deposits within the Central Ebro basin, Northest Spain. Land Subsidence, IAHS Publication No 200, pp 323 - 332.

Berg, W. R. (1980): Subsidence Due To Underground Mining In The Hanna Coalfield. Master's Thesis, Univ. of Wyoming, Laramic, Woyming.

Bergado, D.T., Balasubramanium, A.S. and Apaipong, W. (1984): Field monitoring of subsidence effects in AIT campus, Bangkok, Thailand. Land Subsidence, IAHS Publication No 151, pp 391 - 404.

Berry, D. S. (1963): Ground movement considered as an clastic phenomenon. The Mining Engrs., v 37, pp 28 - 39.

Berry, D. S. (1964): A theoretical elastic model of the complete region affected by mining a thin seam. Prec. 6th Symp. Rock Mech., Rolla, Mo., pp 310 - 329.

Berry, D.S. (1977): Progress in the analysis of ground movements due to mining. Large Ground Movements and Structures, Pentech, London, pp 781 - 811.

Berry, D. S. (1964): A discussion of the stochastic theory of ground movement. Rock Mechanics and Engineering Geology, pp 213 - 227.

Berry, D. S. and Sales, T. W. (1960): An elastic treatment of ground movement due to mining. Part I, J. Mech. Phys. Solids, pp 280 - 292

Berry, D. S. and Sales, T. W. (1961): An elastic treatment of ground movement due to mining. Part 2, J. Mech. Phys. Solids, pp 52 - 62.

Berry, D. S. and Sales, T. W. (1962): An elastic treatment of ground movement due to mining. Part 3, J. Mech. Phys. Solids, pp 73 - 83.

Berry, D.S. (1964): The ground considered as a transversly isotropic material.Int. J. Rock Mech. Mining Sci., pp 159 - 167.

Bhattacharya, A. K., Shu, D. M. and Lama, R. D. (1988): Prediction of subsidence effects on a service device from the partial extraction of an underground protective pillar, Proc. of the Conference on Building and Sturctures Subjected to Mine Subsidence, Maitland, pp 72 - 78.

Bhattacharya, A.K. and Shu, D. M. (1989): Mathematical modelling of surface subsidence in the coalfields of New South Wales using a back analysis technique, Land Subsidence, Oxford and IBH Pub., New Delhi, pp 91 - 102.

Bischke, R. E. and Gelty, P.S. (1984): A method for assessing the potential of mine subsidence at abandoned mine site through the assistance of finite element. Proc. of the 24th. US Symp. on Rock Mechanics, Evanston, Illinois, pp 722 - 726.

Bitelli, G., and Russo, P. (1991): Levelling data management for the monitoring of land subsidences. Land Subsidence, IAHS Publication No. 200, pp 453 - 462.

Bles, J. L. and Feuga, B. (1986): The Fracture of Rocks. North Oxford Academic Publishers Ltd., pp 1 - 131.

Bock, O., Thom, C., Kasser, M. and Fourmaintraux, D. (1995) : Development of a new airborne laser subsidence measurement system, aiming at mm-accuracy. Land Subsidence, IAHS Publication No 234, pp.113-122.

Bodziony, J., Litwiniszyn, J. and Smolarski. A, (1960): New Research Into Rock Masses Treated as Media Characterized by Stochastic Equations. Proc. Int. Conf. Strata Control, Paris, pp 137 - 150.

Boling, J. K. (1984): Earth fissure movements in South Central Arizona, USA. Land Subsidence, IAHS publication No 151, pp 757 - 766.

Bortolami, G., Carbognin, L. and Gatto, P. (1984): The natural subsidence in the lagoon of Venice, Italy. Land Subsidence, IAHS Publication No 151, pp 777 - 784.

Bott, M. H. P. (1976): Mechanisms of basin subsidence - an introductory review. Tectonophysics, pp 1 - 4.

Brackley, I. J. A., Rosewarne, P. N. and Grady, L. A. (1984): A prediction of surface subsidence caused by lowering the water table in Dolomite. Land Subsidence, IAHS Publication No 151, pp 127 - 136.

Barends, F. B. J. (1989): Man - induced indirect land- subsidence of alluvial planes. Land Subsidence, Oxford and IBH Pub., New Delhi, pp 27 - 34.

Brauener, G. (1973a): Calculation of ground movement in European coalfields. Proc. Symp. Subsidence in Mines, Australian Inst Min. Met., pp 10/1 - 10/8.

Brauener, G. (1973b): Subsidence Due to Underground Mining ; Part I, Theory and Practice in Predicting Surface Deformation.U.S. Dept. of Interior, Bureau of Minis, 56 p.

Brauener,G. (1973c): Subsidence due to Underground Mining, Part 2; Ground Movement and Mining Damage. Report No. 1C 8571 and 8572, Washington, DC: USBM. 56 p and 53 p.

Bravo, R., Rogers, J. R. and Cleveland, T. G. (1991): Analysis of ground water level fluctuations and borehole extensometer data from the Baytown area, Haustom, Texas, Land Subsidence, IAHS Publication No 200, pp 655 - 666.

Bravo, R., Rogers, J. R. and Cleveland, T. G. (1991): On the determination of the compressible soil properties required to model subsidence in the area of Houston, Texas. Land Subsidence, IAHS Publication No 200, pp 667 - 676.

Bravon, A. (1986): Land Use Recommendation to Minimise Subsidence Risk in Superior, Woyming. Report to Woyming Dept of Env. Quality, Midwest Mining Company, Littleton Co., Report 1072/06

Briggs, H. (1929): Mining Subsidence. Arnold & Co., London, 215 p

Brighenti, G. (1991): Land subsidence due to thermal water withdrawal : the case of Abano Terme, Northern Italy. Land Subsidence, IAHS Pbulication No 200, pp 515 - 524.

Brighenti, G. and Fabbri, S. (1984): Criteria for evaluating soil compressibility and permeability in areas subjected to subsidence due to hydrocarbon reservoirs production. Land Subsidence, IAHS Publication No 151, pp 641 - 650.

Brook, D. (1991): Abandoned limestone mines in the West Midlands of England - a strategy for action. Land Subsidence, IAHS Publication No 200, pp 215 - 224.

Brook, G.A. and Allison, T.L. (1984): Fracture mapping and ground subsidence susceptibility modelling in covered karst terrain : The example of Dougherty County, Georgia. Land Subsidence, IAHS Publication No 151, pp 595 - 606.

Bruhn, R. W., Magnuson, M. O. and Gray, R. E. (1978): Subsidence over mined out Pittsburgh Coal. ASCE Convention Coal Mine Subsidence Session, Pittsburgh PA, USA, pp 26 - 55

Bruhn, R. W., Speek, R. C. and Thill, R.E. (1983): The Appalachian Field; Damage to structures above active underground mines. Surface Mining, Environmental Monitoring and Reclamation Handbook, Elsevier Sc.Pub.Co., Inc.New York, pp 657 - 669.

Bruhn, R. W. and Speek, R. C. (1983): The Appalachian Field: A surface monitoring program over pillar extraction mine panels. Surface Mining Environmental Monitoring & Reclamation Handbook, pp 647 - 656.

Bruhn, R. W., Magnuson, M. O. and Gray, R. E. (1980): Subsidence over abandoned mines in the Pittsburgh coal bed. Proc. 2nd. Gr. Movements and Structures Pentech Press, London, pp 142 - 156.

Bruhn, R.W., Speck, R. C., McCann, W. S. and Cecil, O. (1983): Survey of Ground Surface Conditions Affecting Structural Responce to Subsidence. Contract Report, US Bureau of Mines Contract No. J0295014. Monroeville, PA: GAI Consultant, Inc.

Buchheim, W. (1966): The mathematical treatment of subsidence processes in loose masses as a process of diffusion of rock gaps. Proc. Symp. Rock Mech., Cracow (Poland). pp 67 - 74

Bukrinskogo, B. A., et.al. (1984): Subsidence of rock and surface due to underground mining. Moske Mining, pp 104 - 106.

Bull,W.B. (1961): Causes and Mechanics of Near-surface Subsidence in Western Fresno Country, California. USGS Professional Paper 424 - B: US Govt. Printing office, Washington DC. pp 187 - 189.

Bull, W. B. (1963): Alluvial Fans and Near Surface Subsidence in Western Fresno Country, California. USGS Professional Paper 437 - A: US Govt. Printing office, Washington DC. 70 p.

Bull, W. B. (1974): Geological factors affecting compaction of deposits in a land subsidence area. Geol. Soc. of Amer. Bull.84, pp 3783 - 3802.

Bull, W. B. (1975): Land Subsidence Due to Groundwater Withdrawal in the Lon Banos-kettleman City Area, California, Part 2, Subsidence and Compaction of Deposits.U.S. Geological Survey Professional Paper, 437-F, 90 p.

Burland, J.B. and Worth, C.P. (1975): Settlement of Buildings and Associated Damage. Publication CP33/75, BRE, Watford.

Burland, J.B. (1967): Deformation of Soft Clay, Ph.D.Thesis, Univ. of Cambrige.

Burland, J.B. and Moore, J.F.A. (1973): The measurement of ground displacements around deep excavations. Proc. British Geotech. Sco. Symp. on Field Instrumentation. BSR, Garston, Waford, WD2, 7JR.

Burland, J. B. and Worth, C. P. (1975): Settlement of Buildings and Associated Damage. Review Paper: Session V, Settlement of Structures, John Willy & Sons, New York, pp 611- 654.

Bush, T. W. (1954-55): Statutory requirement in new coal mining areas. Trans. Inst. Min. Engrs., n.114, p 752 .

CANMET (1985): Report on Seminar on Postash Mining and Ground Control. Energy, Mines and Resources Canada, Report MRP/MRL, 85- 1, 75 p.

CMPDIL (RI-1) (1987): Report on Preliminery Investigation of Reclamation of Surface Subsidence Caused by Unerground Coal Mining in Ranigunj Coalfield, (Unpublished).

Cacon, S. and Deeb, F. (1995) : The influence of water reservoirs on crustal deformations in an active tectonic trough. Land Subsidence, IAHS Publication, No.234, pp.123-128.

Cameron - Clarke, I. S. (1986): The distribution and nature of surface features related to shallow undermining on the East Rand. Proc. SANGORM, 1986 Symp., pp 33 - 38.

Cancelli, A. (1984): Geotechnical properties of Modena subsoil : A preliminary report. Land Subsidence, IAHS Publication No.151, pp 337 - 346.

Capra, A., Folloni, G. and Russo, P. (1991): Subsidence control in the town of Bologna. Land Subsidence, IAHS Publication No. 200, pp 443 - 452.

Caputo, M. Folloni, G. et.al. (1972): Survey and Geometric Analysis of Subsidence in the Region of Venice and its Hinterland,CNR - Lab. Studio Dinamica Grandi Masse, Venezia, TR 9.

Carnec, C., Kding, C. and Massonnet, D. (1995) : Measurment of land subsidence by means of differential S.A.R. interferometry. Land Subsidence, IAHS Publication, No.234, pp139-148.

Carbognin, L., Tosi, L. and Teatini, P. (1995) : Analysis of actual land subsidence in Venice and its hinterland (Italy). Land Subsidence, IAHS Publication, No.234, pp129-138.

Carbognin, L. and Gatto, P.(1984): An overview of the subsidence in Venice. Land Subsidence, IAHS publication No. 151, pp 321 - 328.

Carbognin, L. and Poland, J.F. (1984): Highlights of groundwater withdrawal sessions. Land Subsidence, IAHS Publication No.151, pp 931-932.

Carbognin, L., Gatto, P. and Marabini, F. (1984a): The City and the Lagoon of Venice, A Guidebook on the Environment and Land Subsidence. Ufficio Stampa, Comunicazione e Informazione, Comum di Moderea, Italy. 36 p.

Carbognin, L., Gatto, P. and Marabini, F.(1984b): Correlation between shoreline variation and subsidence in the Po river delta, Italy. Land Subsidence, IAHS Publication No. 151, pp 367 - 372.

Carbognin, L., Gatto, P., Moozi, G. et.al. (1976): New trend in the subsidence of Venice. Land Subsidence, IAHS Publication No. 121, pp 65 - 81.

Carpenter, M.C. and Bradley, M.D. (1984): Legal perspectives on subsidence caused by ground water withdrawal in Texas, California and Arizona, USA. Land Subsidence, IAHS Publication No. 151, pp 817- 828.

Carree, G. J. and Hulsbergen, J.G. (1984): A method for estimating building damage in subsiding areas. Land Subsidence, IAHS Publication No. 151, pp 907 - 914.

Carrera, G. H. and Vanicek, P. (1984): Review of techniques for determining vertical ground movements from levelling data. Land Subsidence, IAHS Publication No. 151, pp 195 - 202.

Carter, P. (1989): Mining subsidence in Scotland - A review. Land Subsidence, Oxford & IBH Pub., New Delhi, pp 245 - 258.

Carter, P.G. and Johnson, A.I. (1989): Land subsidence - The last five years: Report by Technical committee TC - 12. Proc. 12th. Int. Conf. on Soil Mech. and Foundation Engg., Rio de Janerio, pp 3201 - 3205.

Carter, P.G. and Winning, J.M. (1988): Mavisbank house, structural protection against mining subsidence. Proc. 2nd. Int. Conf. on Construction in Areas of Abandoned Mine Workings, Edinburgh, pp 191 - 197.

Carter, P.G. et.al. (1980): Mining subsidence in Bathgate. A town study. Proc. 2nd. Conf. Gr. Movements & Structures, Cardiff.

Carter, P.G. et.al. (1980): Case histories which break the rules. Proc. Int. Conf. on Construction in Areas of Abandoned Mine Workings. Edinburgh. pp 20 - 30.

Carzera, G. and Vanicek, P. (1988): Compilation of a new recent vertical crustal movements map for Canada. Proc. CONFERENCE 1988, Fredericton, New Burnswick, Canada, pp 113 - 118.

Chaffins, R.D. (1987): Sub-surface Strata Movements due to Longwall Mining. M.S. Thesis, Dept. of Min. Engg., West Virginia University, Morgantown, WV, USA.

Chakraborty, A. K. and Singh, K. B. (1989): Effects of mining subsidence on environment and ecology - the remedies. Land Subsidence, Oxford & IBH Pub., New Delhi, pp 575 - 587.

Chatterjee, D.S. and Saxena, N. C. (1996): Subsidence prediction methodologies in India. MINETECH, v 17, n 1, pp 62-69.

Chattopadhyay, S. and Bhattacharaya, J. (1989): Visibility analysis and landscape design of sites damaged due to mining. Land Subsidence, Oxford & IBH Pub., New Delhi. pp 531 - 544.

Chen, C. Y. (1983): Subsidence control measures. Min. Engrs.

Chen, D.W. and Peng, S.S. (1981): Analysis of Surface Subsidence Parameters due to Underground Longwall Mining in the Northern Appalachian Coalfield. Report No. TR 81-1. Morgantown, West Virginia University.

Chen, Jie Jia. (1987): Application of the fuzzy similar selection principle to the study of land subsidence. Urban Geology of Coastal Lowlands in China, Atlas of Urban Geology, 3.

Chen, Yong-gi. (1988): Deformation surveys in P. R. Chuia. Proc. CONFERENCE 1988, Fredericton, New Burnswick,Canada, pp 136 - 141.

Chen, J. and Xiang, S. (1991): Sinkhole collapse resulting from pumping of karst ground water : a problem and its solution. Land Subsidence, IAHS Publication No 200, pp 313 - 322.

Chen, Y.Q. (1983): Analysis of Deformation Surveys - A Generalised Approach. Ph. D. dissertation, Univ. of New Burnswick, Canada.

Chen, Y. Q. and Chrzanowski, A. (1982): A general approach to the interpretation of deformation measurements. Proc. of the Centennial Meeting of CIS, Ottawa, v 2, pp 247 - 266.

Cherny, G.I. (1966) : Determination of surface subsidence and deformation of rock movement in the form of rheological flow. Izvestiga VUZ. Gorny Zhurnal. N 9, pp 3-9.

Chilingarian, G.V., Katz, S.A. and Khilyuk, L.F. (1996): Neural network based prediction of subsidence, surface gas migration and seismic activity. Land Subsidence, IAHS Publication No.234, pp 41-46.

Chi, S. C. and Reilinger, R. E. (1982): Geodetic evidence for subsidence due to ground water withdrawal in many parts of the U.S., J. of Hydrology, n.36.

Choi, S. K. and Coulthard, M. A. (1990): Modelling of jointed rockmass using the district element method. Proc. Int. Conf. on Jointed and Faulted Rock, Vieana.

Chrzanowski, A., Cheu, Y. Q., Leal, J., Murria, J. and Paplawski, T. (1991): Use of global positioning system (GPS) for ground subsidence measurements in Western Venezuela oil fields. Land Subsidence, IAHS Publication No. 200, pp 419 - 432.

Chrzanowski, A., Faig, W., Kurz, B. and Makosinski, A. (1980): Development, Installation and Operation of a Subsidence Monitoring and Telemetry System. Contract Report to Canada Centre for Mineral and Energy Technology, Calgary.

Chrzanowski, A. S. and Chrzanowski, A. (1991): Modelling and prediction of ground subsidence using an iterative finite element method. Land Subsidence, IAHS Publication No 200, pp 173 - 180.

Chrzanowski, A. and Chrzanowski, A. S. (1984): A comparison of empirical and deterministic prediction of mining subsidence. Land Subsidence, IAHS Publication No. 151, pp 137 - 146.

Chrzanowski, A. and Hart, F. (1983): Role of the survey in North America in studies of ground movements in mining areas. Proc. of the ACSM Annual Convention, Washington, D.C., pp 262 - 271.

Chrzanowski, A., Cheu, Y. Q. and Second, J. (1988): On the strain analysis of tectonic movements using fault crossing geodetic surveys. Tectonophysics, v 97, pp 297 - 315.

Chrzanowski, A., Cheu, Y. Q., Leeman, R. and Leal, J. (1988): Integration of the global positioning system with geodetic levelling surveys in ground subsidence studies. Proc. CONFERENCE 1988, Ferdericton, New Burnswick, Canada, pp 156 - 169.

Chrzanowski, A., Faig, W., Kurz, B. and Makosinski, A. (1983): Maintenance, Construction and Installation of an Additional System, and Research on Further Improvements and Extended Applications of the Subsidence Monitoring and Telemetry System. Contract Report to Canada Centre for Mineral and Energy Technology, Calgary.

Chrzanowski, A. and Szostak-Chrzanowski, A. (1985): Finite element modelling ground movement over a steepy inclined coal seam. Proc. VI ISM Symp. Harrogate, England, pp 709 - 718.

Chrzanowski, A. and Chen, Y. Q. (1986): Report of the Ad Hoc committee on the analysis of deformation surveys. Proc. XVIII Int. Cong. FIG, Toronto, Paper No. 608.1.

Chugh,Y.P. (1989): Subsidence research: Where do we go from here. Land Subsidence, Oxford & IBH Pub., New Delhi, pp 45 - 50.

Chugh, Y.P., Hao, Q.W. and Zhu, F.S. (1989): State-of-the- art on mine subsidence prediction, Land Subsidence, Oxford & IBH Pub., New Delhi. pp 51 - 66.

Chugh,Y.P., Kusan, S., Atri, A. and Sweigard, R. (1988): Subsidence monitoring at a shallow partial extraction room and pillar mine in Midwestern United States. Proc. of the 7th. Int. Cong. on Mine Surveying, USSR, Leningrad, pp 21 - 33.

Chugh,Y.P. and Atri, A. (1989): Subsidence monitoring at a shallow partial extration room and pillar mine in Midwestern United States. Proc. 30th. US. Symp. on Rock Mech., pp 715 - 722.

Chugh, Y. P., et. al. (1986): An analysis of subsidence movements associated with an abandond shallow room and pillar coal mines. Proc. 2nd. Work. on surface Subsidence due to Underground Mining, Morgantown, WV, USA, pp 194 - 203.

Cifelli, R. C. and Rauch, H. W. (1986): Dewatering effects from selected underground coal mines in North - Central West Verginia. Proc. 2nd. Work. Surf. Subsidence due to Underground Mining, Morgantown, WV, USA, pp 249 - 263.

Cividini, A. et. al. (1881): Some aspects of characterization problems in geomechanics. Int. J. Rock Mech. & Min. Sci. & Geomech. Abstr., v 18, pp 487 - 503.

Claessen, F. A. M. (1984): An integrated study to forecast and to present detrimental effects in the province of north Holland resulting from a change in the groundwater regime after the reclamation of the Markerwaard Polder. Land Subsidence, IAHS Publication No 151, pp 869 - 876.

Clarke, A. M. (1962): Some structural, hydrological and safety aspects of recent developments in South-East Denhum. The Mining Engineer.

Colombetti, A., Gelmini, R., Pellegrini, M., Paltrinieri, N. and Zavatti, A. (1984): Land subsidence in the the area of modena, Po Valley, Nothern Italy. Land Subsidence, IAHS Publication No 151. pp 805 - 816.

Comptroller General (1979): Alternatives to Protect Property Owners From Damages Caused by Mine Subsidence. Report No. CED 72 - 25. Washington, DC. US General Accounting Office.

Conroy, P. J. and Gyarmaty, J. H. (1982): Planning subsidence monitoring programs over longwall panels. Proc. Conf. on State of the Art of Ground Control in Longwall Mining and Mine Subsidence, Soc. Min. Eng. Am. Inst. Min. Met. & Pet. Eng, New York, USA, pp 225 - 234.

Construction Over Abandoned Mine Working (1984): Healy & Head CIRIA Special Publication No. 22.

Contaldo, G.J. and Mueller, J. E. (1991): Earth fisseres and land subsidence of the Mimbres basin, Southwestern New Maxico, USA. Land Subsidence, IAHS Publication No 200. pp 301 - 310.

Coprapcioglu, M.Y., Bear, J. and Mathur, S. (1984): One dimensional land subsidence with variable total stress. Land Subsidence, IAHS Publication No 151. pp 19 - 28.

Corapcioglu, M. Y. and Bear, J. (1983): A mathematical model for regional land subsidence due to pumping: 3 Integrated Equations for a Phreatic Aquifer. Water Resource Research, v 19, n 4, pp 895 - 908.

Corden, C. H. H. (1964): The Recording of Boundary Distortion in Mining Subsidence. Ph. D. Thesis, Univ. of Leeds, UK.,146p.

Corden, C. H. H. and King, H. J. (1965): A field study of the development of surface subsidence. Int. J. Rock Mech. Min. Sci., v 2, n1, pp 43 - 55.

Cotecchia, V. (1984): Subsidence phenomenon due to earthquakes, Italian cases. Land Subsidence, IAHS Pbulication No 151. pp 829 - 840.

Coulthard, M.A. and Dutton, A.J. (1988): Numerical modelling of subsidence induced by underground coal mining. Proc. 29th. US. Symp. on Rock Mech., pp 529 - 536.

Crang, W.R. (1931): Essential Factors Influencing Subsidence and Ground Movement. Internal Circular 6501, USBM, 14 p.

Crouch, S.L. (1976): Analysis of Stresses and Displacements Around Underground Excavation: An Application of the Displacement-Discontinuity Method. Geomech. Report to the National Sc. Foundation, Minniapolis, Univ. of Minnesota, 268p.

Crouch, S.L. (1983): Boundary Element Methods in Solid Mechanics with Application in Rock Mechanics and Geological Engineering. George Allen and Unwin, UK., 322p.

Cundale, P.A. (1980): UDEC-A Generalised District Element Program for Modelling Jointed Rock. US Army Europ. Res. Off. and Def. Nucl. Agency Contract Data, 37-79-C-0548.

Cundall, P.A. (1972): A computer model for simulating prograssive large scale movements in blocky rock system. Proc. of the Symp. of ISRM, 1, Nomcy, France.

Cunietti, M., Facchinelli, F., Mozzi, G., Ricceri, G., Soranzo, M. and Vitelli, E. (1984): An experimental investigation on the stability of levelling bench marks inserted into sedimentary soils. Land Subsidence, IAHS Publication No 151, pp 203 - 212.

DOE (1976): Reclamation of Derelict Land : Procedure for Locating Abandoned Mine Shafts, Planning, Regional and Minerals Directorate, Dept. of Env., London, 88 p.

DOE (1983): Limestone Mines in the West Midlands : the Legacy of Mines Long Abandoned. Dept. of Env. London, 24 p.

Daeman, J.J.K. and Hood, M. (1991): Subsidence profile functions derived from mechanistic rock models. Proc. of the Workshop on Surface Sub sidence due to Undergroung Mining, Margantown, WV, USA, 28p.

Dahl, H.D. (1972): Two and three dimensional elastic- elastoplastic analysis of mine subsidence. Proc. of the 5th. Int. St. Cont. Conf., London.

Dahl, H. D. and Choi, D. S. (1973): Some case studies of mine subsidence and its mathematical modelling. Proc. of the 15th. US Symp. on Rock Mech., South Dakota, pp 1 - 21.

Daito, K., Mizuno, M. and Ueshita, K. (1991): Control of groundwater withdrawal for preventing land subsidence in the Owari Plain, Japan. Land Subsidence, IAHS Publication No 200, pp 533 - 542.

Dassargues, A. and Li, X.L. (1991): Computing the land subsidence of Shanghai by a finite element method. Land Subsidence, IAHS Publication No 200, pp 613 - 624.

Davies, J. D. (1960): Circular tanks on ground subject to mining subsidence. Civil Eng. and Pub. Works Rev.,v 55,n 648, pp 918 - 920.

Dazzi, R., Gatto, G., Moozi, G. and Zambon, G. (1989): Ground evaluation of Venice and its evolution. Land Subsidence, Oxford & IBH Pub., New Delhi, pp 271 - 282.

DeRossett, C. and Unrug, K.F. (1993): Surface fracture development from mine subsidence in Eastern Kentuky, Proc. 12th. Int. Conf. Gr. Cont. in Min., Morgantown,WV, USA, pp 397 - 404.

Dean, J.N. (1967): Old mine shafts and their hazards. The Mining Engineer, v 126, pp 368 - 376.

Degirmenci, N., Reddish, D.J. and Whittaker, B.N. (1988): A Study of Surface Subsidence Behavior Arising from Longwall Mining of Steeply Pitching Coal Seams. 6th.Coal Congress, Zonguldak, Turkey, 27p.

Dejean, M.J.P. and Martin, F. (1973): Amplitude of subsidence of underground openings subject to influence of mining adjacent and below. Proc. Symp. Subsidence in Mines, Aust. Inst. Min. Met.,pp 4/1 - 4/5.

Dept. Of Environment, (1984): Treatment of Disused Mine Openings. H.M.S.O., London.

de Glopper, R. J. (1984): Subsidence in the recently reclaimed Ijsselmeerpolder "Flevoland". Land Subsidence, IAHS Publication No 151, pp 487 - 496.

de Glopper, R.J. and Viergever, M.A. (1984): Prolonged subsidence observations in soft sediments under a sand load. Land Subsidence, IAHS Publication No 151. pp 445 - 454.

Dhar, B. B., Saxena, N. C. and Singh, U. K. (1992): Rockmass characterization for estimation of support load in underground galleries and prediction of surface subsidence in Indian coalfields. Eurok '92, Proc. on Rock Charaterization.

Dhawan, Y.P. and Singh, P.N. (1989): Subsidence problems in TISCO collieries, Jharia. Land Subsidence, Oxford & IBH Pub., New Delhi, pp 441 - 454.

Diamond, J. D. and Williamson, A. K. (1983): A Summary of Groundwater Pumpage in the Central Valley, California. 1961 - 1977. US Geological Survey Water Resources Investigations Report 83-4037, 70p.

Dickinson, J. (1898): Subsidence due to colliery workings. Proc. Min. Geol. Soc., v 25, p 600.

Dixon, J.S. (1885): Some note on subsidence and draw. Trans. Min. Inst. Scotland, v 7, 224 p.

Donaldson, E.C. (Editor) (1983): Proceedings of the 1982 Forum on Subsidence due to Fluid Withdrawl, US Dept. of Energy and Bartlesville Energy Technology Center, Bart -lesville, OK, USA, 141p.

Donovan Jacobs, J. (1975): Some tunnel failures and what they have taught. Hazards in Tunnelling and on Falsework, ICE, London, pp 37 - 46.

Dragert, H., Georgian, Y. and Klensberg, A. (1988): Establishment of crustal deformation network using GPS.A case study. CISM .J. ACSGC, G 2, n 4, pp 341 - 351.

Dudley, J.H. (1970): Review of collapsing soil. J. Soil Mech. & Foundation Division, v 96, n SM-3, pp 925 - 947.

Dumont, J.F. and Garcia, F. (1991): Active subsidence controlled by basement structures in Maranon basin in Northeastern Peru. Land Subsidence, IAHS Publication No 200, pp 343 - 350.

Dunrud, C.R. (1976): Some Engineering Geologic Factors Controlling Coal Mine Subsidence in Utah and Colorado. US Geological Survey Professional Paper, v 969, 39p.

Dunrud, C.R. (1984): Coal mine subsidence, Western United States. Man Induced Land Subsidence, Revs. in Engg. Geol. The Geo.Soc.of Amer., v 6, pp 151 - 194

Dunrud, C. R. and Osterwald, F.W. (1978a): Effects of Coal Mine Subsidence in the Western Powder Basin,Wyoming. US Geological Survey open file Report, 48-473, 71p.

Dunrud, C. R. and Osterwald, F. W. (1980): Effects of Coal Mine Subsidence in the Sheridon, Wyoming Area. Geological Survey Professional Paper 1164.

Dunrud, C.R. and Osterwald, F.W. (1987): Coal mine subsidence near Sheriday, Wyoming. Bull. Ass.Engg. Geologists, XV(2), pp 175 - 190.

Eckardt, A. (1913): Die mechanischen einwirkungen des Abbaus auf das Verhalten des Gebriges. Gluckauf, v 49, pp 353-361, 397- 403.

Edwards, J.L. and Swan, D.B. (1993): Mine design considerations for surface subsidence control. Proc. 12th. Int. Conf. on Gr. Cont. in Mining, Morgantown, WV, USA, pp 385 - 396.

Ege, J.R. (1984): Mechanisms of surface subsidence resulting from solution extraction of salt. Man-Induced Land Subsidence, Review of Eng. Geology, The Geological Soc. of America., pp 203 - 221.

Eggelsmann, R.F.(1984): Subsidence of peat land caused by drainage, evaporation and oxidation. Land Subsidence, IAHS Publication No 151, pp 497 - 506.

Egger, K. (1988): Dam monitoring and geodetic deformation measurements. Proc. CONFERENCE 1988, Fredericton, New Burnswisk, Canada, pp 170 - 174.

Ehrhardt, W. and Sauer, A. (1961): Pre-calculation of subsidence, tilt, and curvature over extractions in flat formations. Bergbauwissen-schaften, v 8, pp 415 - 428.

Elifrits, C.D. and Aughenbaugh, N.B. (1986): Effects of moisture variation and overburden on subsidence. Min. Engg. J., v 38, n 2, pp 115 - 119.

Esaki, T., Dohzono, S., Kimura, T. and Kameda, N. (1991): Environmental system for subsidence engineering. Land Subsidence, IAHS Publication No 200, pp 163 - 172.

Esaki, T., Shikata, K., Aoki, K. and Kimura, T. (1991): Surface subsidence in natural gas fields. Land Subsidence, IAHS Publication No 200, pp 109 - 118.

Ewy, R. T. and Hoodd, M. (1984): Technical Note - Surface strain over longwall coal mines : Its relation to the subsidence trough, curvature and to surface topography. Int. J. Rock Mech. Min. Sci. & Geomech. Abstr., v 21, n 3, pp 15 - 19.

Eysink, W.D., Hug, P., Dijkema, K.S., Smit, C.J. and Dobben, H.J.van, (1995) : Monitoring the effects of subsidence of the coastal island Ameland in the Wadden Sea. Land Subsidence, IAHS Publication, No. 234, pp 259-268.

Faig, W. and Armenakis, C. (1982): Subsidence monitoring by photogrammetry. Proc. 4th canadian Symp. Min. Sur. & Def. Meas., Banff.

Farmer, I. W. and Altounyan, P. F.R. (1980): The mechanics of ground deformations above a caving longwall face. Proc. 2nd. Int. Symp. on Gr. Movements and Structures, Cardiff, pp 75 - 91.

Farran, C. E. (1954): The effect of mining subsidence on land drainage. J. Inst. Engrs.,Wat. & Wat.Engrs., v 58, p 197.

Fejes, A. J. (1985): Surface subsidence resulting from longwall mining in central Utah. Proc. 26th Symp. on Rock Mech., p 197.

Fisekci, M. Y. and Chrzanowski, A. (1981): Some aspects of subsidence monitoring in different terrain and climate conditions of rocky mountains, Western Canada. Proc. Workshop on Sur. Subsidence due to underground Mining, Morgantown, WV, USA, pp 218 - 238.

Fisekci, M.Y., Chrzanowski, A., Das, D.M. and Larocque, G. (1981): Subsidence studies in thick and steep coal seam mining. Proc. Conf. Gr. Cont. in Mining, Morgantown, WV, USA.

Fitzpatrick, D.J.(1987): Modelling of Mining Subsidence Mechanisms and Prediction of Ground Movements. Ph. D. Thesis, Univ. of Nottingham, UK.

Flaschentragen, H. (1957): Considerations in ground movement phenomena based on observations made in the Left Bank Lower Rhine Region. Proc. Europ. Cong. Ground Movement, Leeds, pp 58 - 73.

Flouzat, M., Fourmaintraux, D. and Camphuysen, R. (1995) : Advanced continuous monitoring of subsidence above gas fields using spatial geodetic measurements. Land Subsidence, IAHS Publication, No.234, pp.269-280.

Fokker, P.A., Urai, J.L. and Steeneken, P.V. (1995) : Production induced convergence of solution mined caverns in magnesium salts and associated subsidence. Land Subsidence, IAHS Publication, No.234, pp.281-290.

Fontes, J. Ch. and Bortolami, G. (1972): Subsidence of the Venice area during the past 40,000 years.Nature, v 244, n 5415, pp 339 - 341.

Forde, M. C. et.al. (1984): Mineworkings 84. Proc.Int. Conf. on Constructions in Areas of Abandoned Mine Workings, pp 1 - 272. Edinburgh.

Forde, M.C. et.al. (1988): Mineworkings 88. Proc. Int. Conf. on Constructions in Areas Abandoned Mine Workings, 314p. Edinburgh.

Forrester, D. J. and Whittaker, B.N. (1976): Effects of mining subsidence on colliery spoil heaps. Int. J. Rock Mech. & Min. Sc. and Geomech. Abst., v 13, pp 113-120 - 121-133.

Frankham, B.S. and Mould, G.R. (1980): Mining subsidence in North Wales. Proc. Aust. Inst. Min. Metall., v 275, pp 167 - 179.

Frankham, B. S. and Holla, L. (1984): Mining subsidence and its effects on surface development in the coalfields of New South Wales. Proc. 3rd. Int. Conf. on Gr. Movement and Structure, Cardiff, pp 207 - 222.

Franks, C. A. M. and Geddes, J. D. (1984): A comparative study by numerical modelling of movements on sloping ground due to longwall mining. Proc. 3rd. Int. Conf. on Large Gr. Movements and Structures, Cardiff, pp 377 - 396.

Franks, C. A. M. and Goddes, J. D. (1986): Subsidence on steep slopes due to longwall mining. Int. J. Min. & Geol.Eng., v 4, pp 291 - 301.

GIMS.(1958):The Movement of the Rock Masses and of the Surface in the Main Coalfields of the USSR. General Institute of Mine Surveying (GIMS), Ugletekhirdat, Moscow. 250p.

Gallavresi, F. and Rodio, G. (1984): Soil upheaving by grouting to safeguard zones affected by significant subsidence problems : Its application to Vanice as a peculiar example. Land Subsidence, IAHS Publication No 151. pp 707 - 716.

Galvin, J. M. (1982): Total Extraction of Coal Seams : The Significance and Behavior of Massive Dolerite Sills. Chamber of Mines of South Africa, Research Report, 19/82, 80 p.

Galvin, J. M. (1987): Surface subsidence mechanisms - theory and practice, Part II - Subsidence mechanisms in practice. The Coal Journal, n 17.

Gambardella, F and Mercusa, G.(1984): Land subsidence in the delta area river Po : Damage and repairing work. Land Subsidence, IAHS Publication No 151, pp 309 - 320.

Gamberdella, F., Bortolotto, S. and Zambon, M. (1991): The positioning system GPS for subsidence control of the terminal reach of the Po river. Land Subsidence, IAHS Publication No 200, pp 433 - 442.

Gambolati, G. (1972): A three-dimenssional model to compute land subsidence. Bull. I. A. H. S.

Gambolati, G. (1974): Second order theory of flow in three- dimensional deforming media. Water Resource Research, v 10, pp 1217 - 1228.

Gambolati,G.(1980): Fast solution to finite element flow equation by modified conjugate gradient method and Newton iteration. Int. J. Num. Methods Engg., v 15, pp 661 - 675.

Gambolati, G., Gatto, P. and Freeze, R. A. (1974): Mathematical simulation of the subsidence of Vanice. Results. Water Resource Research, v 10, n 3, pp 563 - 577.

Gambolati, G., Gatto, P. and Freeze, R. A. (1974b): Predictive simulation of the subsidence of Venice. Science, v 183, pp 849 - 851.

Gambolati, G. and Freeze, R.A. (1973): Mathematical simulation of the subsidence of Venice.1.Theory. Water Resources Research, v 9, n 3.

Gambolati, G., Gatto, P. and Ricceri,G.(1984): Land subsidence due to gas-oil removal in layered anisotropic soil by a finite element model. Land Subsidence, IAHS Publication No 151, pp 29 - 42.

Gambolati, G. Ricceri, G., Bertoni, W., Brighenti, G. and Vuillermin, E. (1991): Numerical analysis of land subsidence at Ravenna due to water withdrawal and gas removal. Land Subsidence, IAHS Publication No 200, pp 119 - 128.

Garg, S. K. (1984): Formation compaction associated with thermal cooling in geothermal reservoirs. Land Subsidence, IAHS Publication No 151, pp 117 - 126.

Garmulewicz, J.(1988): Integrating long base line acoustics and GPS to monitor sea floor subsidence due to submarine coal mining. Proc. CONFERENCE 1988, Fredericton, New Burnswick, Canada, pp 184 - 194.

Garner, J. H. (1945): Report on the Effect of Mining (Coal) Subsidence on Sewers and Sewage Disposal Works in the West Riding of Yorkshire. The West Riding of Yorkshire Rivers Board.

Gatto, P. and Carbognin, L.(1981): The lagoon of Venice : natural environmental trend and man-induced modification. Hydrological Sc. Bull., v 26, n 4, pp 378 - 391.

Geddes, J.D. (1977): Large Ground Movements and Structures. Pentech Press, London, Plymouth, 1064 p.

Geddes, J. D. (1980): Ground Movements and Structures. Pentech Press, London, Plymouth, 964 p.

Geddes, J. D. (1984): Ground Movements and Structures. Pentech Press, London, Plymouth, 876 p.

Geddes, J. D. ed. (1978): Large ground movements and structures. Proc. of the 1st. Int. Conf. London.

Geertsma, J. (1966): Problems of rock mechanics in petroleum engineering. Proc. 1st. Cong. Int. Soc. Rock Mech., Lisbon.

Geertsma, J.(1973): Land subsidence above compacting oil and gas reservoirs. J. Petr. Tech., n 25, pp 734 - 744.

Geron, G. F., Gobbetto, W., Lecis, I., Marzetti, N., Romaro, G., Rossi, G. and Salsilli, F. (1984): Ways to prevent the effects of land subsidence on coastal cities, The Vanice case. Land Subsidence, IAHS Publication No. 151, pp 301 - 308.

General Institute of Mining Surveying(1958): The movements of the rock masses and of the surface in the main coalfields of the Soviet Union. Ugletekhizdat, Moscow, pp 250 p.

Geng, D.Y. and Peng, S.S. (1983): Longwall Mining Techniques for Minimising Surface Structural Damage. S.M.E., A.I.M.E. Annual Meeting, Atlanta, Georgia, Preprint No.83-99.

Gentry, D.W. and Able, J.F. (1978): Surface response to longwall coal mining in mountainous terrain. Asson. Engg. Geologists Bull., v 15, n 2, pp 191 - 200.

Gentry, D.W. and Abel, J.F. (1978): Rock mass responce to mining longwall panel 4N, York Canyon mine. Mining Engineering, v 30, n 3, pp 273 - 280.

Ghosh, A.K.(1982): Extraction below surface structure - appraisal of the Jharia coalfield situation. Mine Subsidence, Parijal Mudra. Pub., Dhanbad, pp 95 - 107.

Ghosh, A.K.(1987): Subsidence management in Indian coal industry - The environmental dimensions. Proc. National Workshop on Env. Management of Mining Operation in India, pp 87 - 94.

Ghosh, H. B., Dugar, M. L., Gautam, N. N. and Saxena, N. C.(1988): Pilot trial of hydropneumatic system of blind backfilling at Ramjeebanpur colliery of Eastern Coalfields Limited. Trans. MGMI, India, v 85, n 1, pp 131 - 149.

Ghosh, R. (1989): Environmental impact of subsidence: A case study from Jharia coalfield, Eastern India. Land Subsidence, Oxford & IBH Pub., New Delhi. pp 565 - 574.

Gibson, D.E.E. (1957): Buildings without foundations. J. R. Inst. Brit. Archit., v 65, p 47.

Gibson, R.E. (1958): The progress of consolidation in a clay layer increasing in thickness with time. Geotechnique, v 8, pp 171 - 182.

Gil, H.(1966): The distribution of displacements in horizontal zones being composed of a granular and an elastic medium. Archiwum Gornictwa, v 11, pp 279 - 310.

Gilluly, J. and Grant, V.S. (1949): Subsidence in the large Beach Harbour Area, California. Geological Soc. of America, Bulletin, v 60, pp 461-530.

Girrens, S. P., Anderson, C. A., Benvett. J. G. and Karmer, M. (1981): Numerical prediction of subsidence with coupled geomechanical - hydrological modelling. Proc. Workshop on Surface Subsidence due to Underground Mining, Morgantown, WV, USA.

Gisotti, G. (1991): A case of induced subsidence for extraction of salt by hydrosolution. Land Subsidence, IAHS Publication No 200, pp 235 - 247.

Glover, C. M. H. and Webster, N. E. (1958): The law relating to damage by mining subsidence and its effects on mining practice. Trans. Inst. Min. Engr., v 118, p 75.

Goel, S. C. and Page, C. H. (1982): An empirical method for predicting the probablity of chimney cave occurrance over a mine area. Int. J. Rock Mech. Sci. & Geomech. Abstr., v 19, pp 325 - 337.

Goodman, G. (1980): Computer Modelling of Mining Subsidence Using the Zone Area Method. M. S. Thesis, VIP 4 SU.

Groenewoud, W., Lorenz, G. K., Brouwer, F. J. J. and Molendjk, R.E. (1991) : Geodetic determination of recent land subsidence in the Netherlands. Land Subsidence, IAHS Publication No 200, pp 463 - 472.

Grand, C. (1969): Mining subsidence and the means of permitting limitation of the effects on the surface. Revue de l'Industric Minerale, v 51, pp 35 - 70.

Gray, R.W. et.al.(1978): Study and Analysis of Surface Subsidence Over the Mined Pittsburg Coal Bed. Bureau of Mines Open File Report, OFR, pp 25 - 78.

Griffith, W. and Conner, E.T. (1972): Mining Conditions Under the City of Scranton, Pa. Bull. No. 25. Washington, DC: US Bureau of Mines.

Grigorovich, V. T. et. al. (1965): Surface subsidence during the working of a sequence of seams at the 'Kaiekan' pit of the Noril'sk Coalfield. Soviet Min. Sci., No.2, pp 86 - 93.

Gromley, J. J. et.al. (1986): Case study : The Glenrock, Wyoming subsidence control project. Proc. in Mine Subsidence, St. Louis, MO, p 83.

Gromley, J.T. (1986): Status and Report of Investigation Gleurock Subsidence Control Project A. Gormley Consultants Inc. , Englewood, Co, USA.

Grond, G. J. A. (1951): The precise topographical measurements in coal mine underground works. Proc. Int. Conf. Rock Pressure and Support of Working. Liege, pp 52 - 66.

Grond, G. J. A. (1953): A Critical Analysis of Early and Modern Theories of Mining Subsidence and Ground Control. Published by the Dept. of Mining, Univ. of Leeds, 57p.

Grond, G. J. A. (1957): Ground movements due to mining. Colliery Engg., v 34, pp 157 - 197.

Grond, G. J. A. (1957): Ground movements due to mining with different types of strata at different depths. Proc. European Cong. Ground Movement, pp 115 - 125.

Gu, X. Y., Tsien, S. I., Huang, H. C. and Liu, Y. (1991): Analysis of Shanghai land subsidence. Land Subsidence, IAHS Publication No 200. pp 603 - 612.

Ground Movements and Their Effects on Structures (1984), Surrey University, 441 p.

Ground Subsidence (1989): A. C. Waltham, Blackie, 202 p.

Gutierrez, F. (1995) : Synsedimentary and postsedimentary subsidence due to gypsum karstification and derived hazards (Calatayud

Graben, Iberian Range, Spain). Land Subsidence, IAHS Publication, No.234, pp 47-58.

Gussinklo, J. (1995) : Gas below a low country. Land Subsidence, IAHS Publication, No.234, pp 235-238.

Guaunghan, S., Ge, W. and Shu, C. (1989): Parameter identification method of mine subsidence. Land Subsidence, Oxford & IBH Pub., New Delhi, pp 67 - 74.

Guangxiao, D. and Yiaogi, Z. (1984): Land subsidence in China. Land Subsidence, IAHS Publiaction No 151, pp 405 - 414.

Gubellini, A. and Russo, P.(1984): Some notes on the redefinition of a vertical reference network for study of the subsidence in the eastern part of Po river valley. Land Subsidence, IAHS Publication No 151, pp 247 - 256.

Gubellini, A., Lombardini, G. and Russo, P.(1984): Application of high precision levelling and photogrammetry to the detection of the movements of an architectonic complex produced by subsidence in the town of Bologna. Land Subsidence, IAHS Publication No 151, pp 257 - 269.

Gubler, E., Kahle, H.G. et.al. (1981): Recent crustal movements in Switzerland and their geographical interpretation. Tectonophysics, Amsterdam, v 71, pp 125 - 152.

Gurtunca, R. G. and Bhattacharyya, A. K. (1988): Modelling of surface subsidence in the southern coalfields of New South Wales. Proc. 5th. Australia-New Zealand Conf. on Geomech., Prediction versus Perfermance, Sydney, pp 346 - 350.

Gurtunca, R.G. and Schumann, E.H.R. (1986): Computer stimulation of surface subsidence using a displacement discontinuity method. Proc. SANGROM Symposium : the Effect of Underground Mining on Surface, South Africa, pp 81 - 87.

Hacckett, P.(1959):An elastic analysis of rock movements caused by mining. Trans. Inst. Min. Engrs., v 118, n 7, pp 423 - 433.

Halbaum, H.W.G. (1903): The action, influence and control of the roof in longwall working. Trans. Inst. Min. Engrs., n 27, 211 p.

Halbaum, H.W.G.(1905): The great planes of strain in the absolute roof of mines. Trans. I. Min. E., n 30, 175 p.

Hall, M. and Orchard, R. J. (1963) : Subsidence profile characteristics. Chart. Surv., v 95, n 8, pp 422 - 428.

Hanjo, Y. and Darmawan, P.(1991): Prediction of future subsidence with quantified uncertainty by an inverse analysis procedure. Land Subsidence, IAHS Publication No 200, pp 625 - 634.

Hansbo, S. (1989): Land subsidence in Sweden - cause and effects. Land Subsidence, Oxford & IBH Pub., New Delhi. pp 37 - 44.

Hao, Q. W. and Ma, W. M. (1985a): On the methods of ground subsidence prediction (in Chinese). J. China Inst. Min. Tech., n 2, pp 37 - 42.

Hao, Q. W. and Ma, W. M. (1985b): Basic patterns of surface subsidence for compound type overlying strata (in Chinese). Proc. Int. Symp. Min. Tech. Sci., Xurhou, China, Mining Technology and Science, China coal Industry Publishing House, 1987, pp 119 - 127.

Haren, G.A. and Sargand, S.M.(1988): Method for assessing effects of longwall mining on surface subsidence. Mining Engineering, pp 451 - 454.

Haroon, M. and Athar, M. (1989): Causes of subsidence in highways and suggested remedial measures. Land Subsidence, Oxford & IBH Pub., New Delhi, pp 623 - 632.

Hasenfers, G. (1984): The Prediction of Surface Subsidence Due to Room and Pillar Mining in the Appalachian Coalfield, M.S. Thesis, VIP & SU.

Howell, F.T. (1984): Photogeology of some salt kasrt subsidence, Cheshire, England. Land Subsidence, IAHS Publication No 151, pp 617 - 628.

Haycock, C. and Karmis, M. (1983): Ground Control Mechanisms in Multi-seam Mining. Finel Report, Prepared for the the Office of Mineral Industries, Bureau of Mines, U.S. Dept. of Interior, 328 p.

Hazine, H. I. (1977): A Study of the Surface Strains Produced by Mining. M. Phil. Thesis, Univ. of Nottingham, UK.

He-yuan, Su., Guo-ming, Lin. and Jia-huan, Qian (1991): Probablity analysis of Shanghai shallow layer subsidence. Land Subsidence, IAHS Publication No 200. pp 585 - 592.

He-yuan, Su. (1984):Mechanism of land subsidence and deformation of soil layers in Shanghai. Land Susbsidence, IAHS Publication No 151, pp 425-434.

Healey, P. R. and Head, J. M. (1984): Construction Over Abandoned Mine Workings. CIRIA special Publication 32, London.

Heasely, K. A. and Saperstein, L.W. (1985): Computer modelling of the surface effects of subsidence control methods. Proc. 26th. US Symp.on Rock Mech., Rapid City, pp 189 - 196.

Heasely, K. A. and Saperstein, L. W. (1986): Practical subsidence prediction for the operating coal mine. Proc. Workshop on Surface Subsidence due to Underground Mining, Morgantown, WV, pp 54 - 67.

Hedley, A.M.(1923): Royal Commission on Mining Subsidence; Minutes of Evidence, H.M.S.O.

Hein, G.W. and Riedl, B. (1995) : A high precision real time land subsidence monitoring system based on DGPS. Land Subsidence, IAHS Publication, No.234, pp 159-168.

Heijmanowski, R. (1995) : Prediction of surface subsidence due to oil or gasfield development. Land Subsidence, IAHS Publication, No.234, pp 291-294.

Hellwell, E. C. (1988): The influence of faulting on ground movement due to coal mining. Proc. 5th. Int. St. Cont. Conf., London, Paper 30, 6p.

Hellwell, E. C. (1988): The influence of faulting on ground movements due to coal mining, the UK and European experience. The Mining Engineer, pp-334 - 337.

Helm, D. C. (1975): One-dimensional simulation of aquifer system compaction near Pixley, California. Water Resource Research, v 11, n 3, pp 465 - 478.

Helm, D.C. (1976): One-dimensional simulation of aquifer system compaction near Pixley, California (II-Stress dependent parameters) Water Resource Research, v 12, n 3, pp 375 - 391.

Helm, D.C.(1977): Estimating parameters of compacting fine grained interbeds within a confined aquifer system by one- dimensional simulation of field observations. Land Subsidence, IAHS Publication No 121, pp 145 - 156..

Helm, D.C. (1978): Field varification of a one-dimentional model for trasient compaction and expansion of a confined auifer system. Proc. Speciality Conf. on Varification of Mathematical and Physical Models in Hydraulic Engineering, College Park, Maryland. American Society of Civil Engineers, pp 189 - 196.

Helm, D.C. (1982): Conceptual aspects of subsidence due to fluid withdrawal. Proc. Int. Conf. Mechanics of Fluids in Porous Media, Nato Inst., Delaware.

Helm, D. C. (1984): Field-based computational techniques for predicting subsidence due to fluid withdrawal. Man Induced Land Subsidence, Rev. Eng. Geology VI, The Geological Soc. of America, pp 1 - 22.

Helm, D. C. (1991): The prediction of heave associated with the storage of surface water in the midst of a region of ongoing subsidence. Land Subsidence, IAHS Publication No 200, pp 545 - 554.

Haneberg, W.C., Reynolds, C.B. and Reynolds, I. B. (1991): Geophysical characterisation of soil deformation associated with earth fissures near San Marcial and Deming, New Mexico. Land Subsidence, IAHS Publication No. 200, pp 271 - 280.

Henley, A.T.(1984): Land subsidence in a California County : The legal issues. Land Subsidence, IAHS Publication No 151, pp 853 - 863.

Heok, E. and Brown, E. T.(1980): Underground Excavations in Rock. London Inst. of Mining and Metallurgy.

Herak, M. and Springfield, V.T.(1972): KARST. Elsevier Publishing Company, Austerdam, 551 p.

Herrera, I., Alberro, J., Grane, R. and Hanel, J. (1976): Development of artificial reservoirs by inducing land subsidence. Land Subsidence, IAHS Publication No. 121, pp 39 - 45.

Highway Research Board (1963): Pavement design in forst areas, II-Design Consideration. Highway Research Board, Record No. 33.

Hiramatsu, Y., Okamura, H. and Sugawan, K. (1979): Surface subsidence and horizontal displacement caused by mining of coal seams. Proc. 4th. Int. Cong. Soc. for Rock Mech., Nontreux, pp 665 - 670.

Hiramatsu,Y. and Oka,Y. (1968): Precalculation of ground movements caused by mining. Int. J. Rock Mech. Min. Sci., n 5, pp 399 - 414.

Hoffman,H.(1964): The effects of direction of working and rate of advance on the scale-deformation of a self-loaded stratified model of a large body of ground. Proc. Int. Conf. Strata Control, New York, pp 397 - 411.

Holdahl, S.R., Zilkoski, D.B. and Holzschuh, J.C. (1991): Subsidence at Houstan, Texas, 1937-1987. Land Subsidence, IAHS Publication No 200, pp3 - 14.

Holla, L. (1985a): Subsidence Investigations in the Southern Coalfields of New South Wales, Australia, Subsidence Report No. 1, Department of Mineral Resources, New South Wales, Sydeny.

Holla, L. (1985b): Mining Subsidence in New South Wales 1, Surface Subsidence Prediction in the Southern Coalfields. NSW Dept. of Mineral Resources, Sydney.

Holla, L. (1985c): Empirical prediction of subsidence movements in the southern coalfields of New South Wales. Proc. 6th. Int. Conf. of Int. Soc. for Mine Surveying, Harrogate, pp 557 - 567.

Holla, L.(1986):Evalution of surface subsidence characteristics in the Newcastle coalfield of New South Wales. The Coal J., n 11, pp 11 - 21.

Holla, L.(1987a): Mining Subsidence in New South Wales, 2, Surface Subsidence in the Newearth Coalfield. Department of Mineral Resources, New South Wales, Sydney.

Holla, L. (1987b): The response of domestic structures to ground movement caused by mining subsidence. Proc. 1st. National Structural Engg. Conf.,Inst. of Engrs. Canberra. pp 472-477.

Holla, L. (1987c):Design of mine working under surface waters in New South Wales. Bull. 14 th. Australasia Instt. Min. Metall., v 292, n 2, pp 45 - 50.

Holla, L. (1988a): Subsidence due to underground coal mining : effects on roads in New South Wales. Proc. 14th. Aust. Road Research Board Conf. Part 5, Australian Raod Research Board, Melborne, pp 11 - 19.

Holla, L. (1988b): Effects of underground mining on domestic sturctures - prediction versus performance. Proc. 5th. Australia - New Zealand Conf. on Geomech., Inst. on Engrs.Canberra, pp 351 - 355.

Holla, L. (1988c): Mining Subsidence and Urban Development. A report prepared by a Ministerial Committee for the Ministers for the Mineral Resources and Minister of Energy, Appendex 3.

Holla, L. (1989): Some aspects of subsidence due to longwall mining in New South Wales, Australia. Land Subsidence, Oxford & IBH Pub., New Delhi, pp 231 - 244.

Holla, L. and Thompson, K. T.(1988): Ground movement in multi- seam mining. Civil Engg. Trans., Inst. of Engrs., Canberra., CE30, n 5, pp 299 - 306.

Holla, L., Hughson, R. A. and Sutherland, G. A. (1986): Subsidence legislation, engineering and management, Australian coal mining practice, Aust. Inst. Min. & Metl., pp 117 - 128.

Holla, L. (1985d): The minimisation of surface subsidence movement by design of mine workings. Proc. Australia's Inst. Min. Metall., no. 290, pp 53 - 59.

Holland, C. T.(1951): What happens and why in multiple-seam mining. Coal Age, v 56, n 8, pp 89 - 93.

Holt, G. E. and Mikula, P. A. (1984): Evalution of Subsidence Prediction Method for Coal Mining in Eastern Australia. Report No. 84-15, Austr. Coal Industry Research Lab, 48 p.

Holzer, T.L.(1984): Ground failure by groundwater withdrawal from unconsolidated sediments - United States. Land Subsidence, IAHS Publication No 151, pp 747 - 756.

Holzer, T.L. (1984): Ground failure induced by groundwater withdrawal from unconsolidated sediments. Man Induced Land Subsidence Rev. Eng. Geology IV, The Geological Soc. of America, pp 67 - 105.

Hood, M., Ewy, R.T. and Riddle, L.R. (1983): Empirical methods of subsidence prediction - a case study of town Illinois. Int. J. Rock Mech. Min. Sci. & Geomech. Abstr., v 20, n 4, pp 153 - 170.

Huck, J.P., Chugh, Y.P. and Jennings, M.(1981): Subsidence control in abandoned coal mines : US practices. Mine Subsidence, Parijat Mudra. Pub. Dhanbad. pp 126 - 135.

Hunt, S. R.(1980): Surface Subsidence Due to Coal Mining in Illinois. Ph. D. Thesis, Univ. of Illinois at Urbana - Champain, 129 p.

I. C. E. (1977): Ground Subsidence. The Inst. of Civil Engrs., London, 99 p.

Iida, K., Sazanami, K., Kuwahara, T. and Ueshita, K. (1976): Subsidence of the Nobi plain. Land Subsidence, IAHS Publication, No.121, pp 47 - 54.

Illinois State Geological Survey.(1980b): Review of Underground Mining Practices in Illinois as Related to Aspects of Mine Subsidence with Recommendations for Legislation. Illinois Institute of National Resources Document No.80/10.

Imam, H. E. (1965): A visco-elastic analysis of mine subsidence. Colliery Eng., v 43, pp 295-301, 343-350.

Imam, H. F.(1965): A Visco-elastic Analysis of Mine Subsidence in Horizontally Laminated Strata. Ph. D. Thesis, Univ. of Minnesota, 126 p.

Ingzam, D.K. (1989): Surface Fractures Development Over Longwall Panels in South Central West Virginia. USBM, RI 9289, 18 p.

Ireland, R. L., Poland, J. F. and Riky, F. S. (1982): Land Subsidence in the San Joaquin Valley Through 1980. U.S. Geological Survey open file report 82-370, 129 p.

Ireland, R. L., Poland, J. F. and Riley, F. S. (1984): Land Subsidence in the San Joaquin Valley, California, as of 1980. U. S.Geological Survey Professional Paper 437-I. pp 143 - 193.

Jain, G. C. (1987): Some case studies - ground and building movements. Proc. 6th. Int. Conf. on Expansive Soil, New Delhi, pp 473 - 476.

Jammal, S.E.(1984): The Winter Park sinkhole and central Florida sinkhole type subsidence. Land Subsidence, IAHS Publication No 151, pp 585 - 594.

Jarosz, A. and Karmis, M.(1986): Control of surface movements above active coal mines in Appalachian. Proc. 2nd. Workshop on Surface Subsidence due to Underground Mining, Morgantown, pp 122 - 133.

Jarosz, A., Karmis, M. and Saroka,A. (1990): Subsidence development time experiences from longwall operations in Appalachian coalfield. Int. J. Min. Geol. Engg., n 8, pp 261 - 273.

Jennings, J. E., Brink, A.B.A., Louw, A. and Grwan, G. D. (1965): Sink-holes and subsidence in the Transvoal dolomites of South Africa. Proc. 6th. Int. Conf. on Soil Mech. & Found. Engg., Montreal, pp 51 - 54.

Jeran,P.W. and Adamek,V. (1988): Subsidence due to undermining of sloping terrian: A case study. Proc. CONFERENCE 1988, Fredericton, New Burnswick, pp 220 - 224.

Jeran, P. W., Adamek, V. and Trevits, M. A. (1986): A subsidence prediction model for longwall mine design. Longwall USA, Pittsberg, Penn., pp 102 - 112.

Jeremic, M.L.(1975): Subsidence problems caused by solution mining of rock-salt deposits. Proc. 10th. Canadian Rock Mech. Symp., Kingston, VI, pp 203 - 233.

Johnson, D. C. (1993): Economic evaluation of subsidence damage mitigation techniques. Proc. 12th. Int. Conf. Gr. Cont. in Min., Morgantown, WV, USA, pp 426 - 431.

Johnson, N.M.(1980): The Relation Between Ephemeral Stream Regime and Earth Fissuring in South Central Arizona, M.S.Thesis, Tueson, Univ. of Arizona, 158 p.

Johnson,W. and Miller, G.C. (1979): Abandoned Coal Mined Lands. US Dept. of Interior, Bureau of Mines, 29 p.

Jones, C.J.F.P. and Ballamy, J.B. (1973): Computer prediction of ground movements due to mining subsidence. Geotechnique, v 23, n 4, pp 515 - 530.

Jones, S. (1986): Pneumatic backfilling : a method for controlling abandoned mine subsidence. Proc. 2nd. Work. on Sur. Subsidence due to Underground Mining, Morgantown, WV, USA.

Jones, T. J. (1945): Strata movements induced by mining. Trans. Inst. Min. Surveyers, v 25, part 1.

Jones, T. J. and Kohli, K. K. (1985): Subsidence over a room and pillar mine in the Appalachian coal province and the use of subsidence perdiction methods - a comparative analysis. Proc. 26th. US Symp. on Rock Mech. Rapid City, pp 176 - 187.

Joshi, R.C. and Horsfield, D.W. (1984): Surface subsidence due to water seepage and the presence of old mine workings in Southern Alberta, Canada. Land Subsidence, IAHS Publication No151, pp 669 - 674.

Kapp, W.A.(1973a):Mine subsidence. Proc. Symp. Subsidence in Mines, Australian Inst. Min. Met., pp 1/1 - 1/9.

Kapp, W.A. (1973b): Subsidence : Kemira colliery, New South Wales. Proc. Symp. Subsidence in Mines Australian Inst. Min. Met., pp 7/1 - 7/9.

Kapp, W. A. (1977): Subsidence investigation in the nothern coalfield, New South Wales and their application to the design of mine layouts in residential area. 11th. Commonwealth Min. & Met. Cong., Hongkong, IMM, London, Paper No.32, 11 p.

Kapp, W.A.(1979): A study of mine subsidence at two collieries in the Sothern Coalfields, New South Wales. Proc. 3rd. Symp. on Min. Sur. & Rock Deformation Measurements, Sudbury, pp 125 - 150.

Kapp, W. A.(1982): Subsidence from deep longwall mining of coal overlain by massive sandstones. Proc. Aust. Inst. Min. Met., v 281, pp 23 - 36.

Kapp, W. A.(1985): Mine subsidence in the Newcastle district, New South Wales. Trans. I. Min. E., Australia, CE 27, n 4, pp 331 - 340.

Karfakis, M. G. (1986): Chimney subsidence - a case study. Proc. 27th. US Symp. on Rock Mech., AIME, Tusealoosa, pp 275 - 282.

Karfakis, M. G. (1987): Mechanism of chimney subsidence over abandoned coal mines. Proc. 6th. Int. Conf. Gr. Cont. in Mining, Morgantown, NU, USA, pp 195 - 203.

Karmis, M.(1989): Predicing ground movements. The Annual Review of Int. Min. Tech. & Development (MINTECH), pp 81 - 84.

Karmis, M. and Haycocks, C. (1983): Computer Simulation of Mining Subsidence Using the Zone Area Method. Dept. Min. & Miner. Eng., VA Politechnic Inst. & State Univ., 62 p.

Karmis, M. and Schilizzi, P. (1984): Development of a subsidence monitoring program for the Southern Appalachian Coalfield. Proc. Int. Conf. at Univ. of Wales, Inst. of Sci. & Tech., Cardiff, pp 223 - 239.

Karmis, M., Agiontantis, Z. and Jarosz, A. (1990): Subsidence prediction techniques in the US, A state-of-the-art review. Min. Res. Eng., v 3, n 3, pp 197 - 210.

Karmis, M., Agiontantis, Z. and Jarosz, A. (1990a): Recent developments in the application of the influence function method for ground movement prediction in the US. Min. Sci. & Tech., n 10, pp 233 - 245.

Karmis, M., Goodman, G. and Eitani,I. (1981): Computer modelling of mining subsidence using the zone area method. Proc. 2nd. US Symp. on Rock Mech., Cambridge, MA, USA, pp 272 - 277.

Karmis, M., Goodman, G. and Hasenfus, G. (1984): Subsidence prediction techniques for longwall and room and pillar panels in Appalachian. Proc. 2nd. Int. Conf. on Stability in Underground Mining, AIME and Univ. of Kentueky, Loxington, USA, pp 541 - 553.

Karmis, M., Haycocks, C. and Triplett, T.(1984): Ground settlement and deformation characteristics above undermined area : experience from the Eastern U. S. coalfield. Proc. Fourth Australian - New Zeland Conf. on Geomech., Parth, pp 647 - 653.

Karmis, M., Haycocks, C., Eitani, I. and Webb, B.(1981): A study of longwall subsidence in the Appalachian coal region using field measurements and computer modelling techniques. Proc. 1st. Conf. on Gr. Cont. in Min., Morgantown, WV, USA, pp 220 - 229.

Karmis, M., Haycocks, C., Webb, B. and Triplett, T. (1982): The potential of the zone area method for mining subsidence prediction in the Appalachian coalfield. Proc. Workshop on Surface Subsidence due to Underground Mining, Morgantown, WV, USA, pp 48 - 59.

Karmis, M., Jarosz, A. and Agiontantis, Z. (1988): An integrated surface movement monitoring system for undermined areas. Proc. 7th.Int. Cong. on Mine Surveying, USSR, Leningrad, pp 42 - 53.

Karmis, M., Jarosz, A. and Schilizzi, P. (1987): Monitoring and prediction of ground movements above underground mines in the eastern U.S.. Proc. 6th. Int. Conf. on Ground Control in Mining, Morgantown, WV, pp 184 - 194.

Karmis, M., Triplett, T. and Schilizzi, P. (1984): Recent developments in subsidence prediction and control for the Eastern U.S. coalfields. Rock Mechanics in Production and Productivity, Proc. 25th. US symp. on Rock Mech., pp 713 - 721.

Karmis, M., Triplett, T., Haycocks, C. and Goodman, G. (1983): Mining subsidence and its prediction in the Applachian Coalfield. Proc. 24th. Symp. on Rock Mech., pp 665 - 675.

Karmis, M., Triplett, T., Schilizzi, P. and Hasenfus, G. (1984): Ground movement characteristics above mined panels in Appalachia - An empirical approach. Proc. Int. Conf. on Case Histories in Geotech Engg., St. Louis, pp 1221 - 1227.

Kartzsch, H. (1963): Ein Optischer Analogrechner zur Vorausbestimmung bergbaulicher Boderbewegungen. Gluckauf, v 99, pp 1033 - 1048.

Kartzsch, H. (1983): Mining Subsidence Engineering. Springer Verlag, 543p.

Kashet, A. I. and Chang, K. R. (1976): Determination of soil subsidence due to well pumping by numerical analysis. Proc. 2nd. Int. Symp. Land Subsidence, Anaheim, California, v 121, pp 167 - 178.

Kauffman, P. W., Hawkins, S.A. and Thompson, R.R. (1981): Room and Pillar Retreat Mining . A Manual for the Coal Industry, Circulation No. 8849, Washington, DC. US Dept. of the Interior.

Kay, S.R.(1898-99): The effect of subsidence due to coal workings upon bridges and other structures. Proc. Inst. Civil Engrs., London, v 135, pp 115 - 123.

Keaton, J.R. and Shlemon, R.J. (1991): The Fort Hancock fissure system, Hudspeth Country, Texas : uncertainties and implications. Land Subsidence, IAHS Publication No 200, pp 281 - 290.

Keinhorst, H. (1934): Consideration of the problems of mining damage. Gluckauf, n 70, pp 149 - 155.

Kelleher, J. T., Van Roosendaal, D.J., Mehnert, B.B., Brutcher, D.F. and Bauer, R.A. (1991): Overburden deformation and hydrologic changes due to longwall coal mine subsidence on the Illinois basin. Land Subsidence, IAHS Publication No 200, pp 195 - 204.

Kendovski, F. S. (1993): Effect of high extraction coal mining on surface and ground waters. Proc. 12th. Conf. Gr. Cont. in Min., Morgantown, WV, USA, pp 412 - 425.

Keorner, R. M. (1984): Acoustic emission monitoring of land subsidence. Land Subsidence, IAHS Publication No 151, pp 225 - 234.

Kersting, N. (1988): Effects of collapsed excavations on ground subsidence in greater distances. Proc. CONFERENCE. 1988, Fredericton, New Burnswick, Canada, pp 240 - 249.

Khair, A. W. (1988): Prediction of ground subsidence in longwall mining areas. Proc. CONFERENCE 1988, Fredericton, New Burnswick, Canada, pp 250 - 259.

Khair, A.W.(1989): Evaluation of monitoring techniques for ground movements over the longwall areas. Land Subsidence, Oxford & IBH Pub., New Delhi. pp 75 - 90.

Khair, A. W. and Molesky, P.J. (1988): Surface ground movements over longwall mining in the Pittsberg seam. Proc. 7th. Int. Conf. Gr. Cont. in Min., Dept. Min. Engg. West Virginia Univ., pp 303 - 308.

Khair, A. W., Chaffins, R. D. and Quinn, M. K. (1987): Study of ground movement over a longwall mine. Proc. 6th. Int. Conf. Gr. Cont. in Mining, Morgantown, WV, USA, pp 141 - 156.

Khair, A. W., Quinn, K. Q. and Chaffins, R. D. (1987): Effect of topography on ground movement due to longwall mining. Min. Engr., v 40, n 8, Feb '87, pp 820 - 822.

King, H.J. (1965): Discussion on ground movement considered as an elastic phenomenon by D. S. Berry. The Mining Engineer, v 40, pp 187 - 188.

King, H. J. and Smith, H. G. (1954): Surface movements due to mining. Colliery Engineering, August,'54.

King, H.J. and Whetton, J.T.(1957): Mechanics of mine subsidence. Proc. Europ. Cong. Gr. Movement, Leeds, pp 27 - 36.

King, H. J. and Whetton, J. T. (1958): Mechanics of mining subsidence, Part I. Coll. Engg., v 35, n 6, pp 247 - 257.

King, H. J. and Whetton, J. T. (1958): Mechanics of mining subsidence, Part II. Coll. Engg., v 35, n 6, pp 285 - 288.

King, H. J., Whittaker, B.N. and Shadbonlt, C.H. (1974): Effects of mining subsidence on surface structures. Mining & The Environment, IMM, London, pp 617 - 642.

Knothe, S. (1953): Rate of advance and ground deformation. Bergerakademic, v 5, n 12, pp 512 - 518.

Knothe, S. (1957): The displacement of the surface under the influence of mining extraction and their interpretation. Proc. Europ. Cong. Gr. Movement, Leeds, England, pp 210 - 218.

Kooij, M.W.A.van der, Halsema, D.van, Groenewoud, W., Ambrosius, B.A., Mets, G.J., Overgaauw, B. and Visser, P.N.A.M. (1995) : Satellite radar measurements for land subsidence detection. Land Subsidence, IAHS Publication, No.234, pp.169-178.

Kochmanski, T. (1959): Integral-Theorie der Bocknbewegunden unter Einwirkung des Abbans (in German). Freib. Forsch. H., A 118, pp 36 - 58.

Kohli, K. K., Peng, S. S. and Thill, R. E. (1980): Subsidence experience in the room and pillar mines of the nothern Applachain coalfield. Proc. Polish - American Conf. on Gr. Cont. in Room & Pillar Mining, SIU, Carbondale, IL.

Koning, H. L. (1984): A detailed case history : Markerwaard, Netherlands, Some introductory remarks. Land Subsidence, IAHS Publication No 151, pp 865 - 868.

Koppula, S. D. (1983): Pore water pressures due to overburden removal. J. Geotech. Eng., ASCE, v 109, n 8, pp 1099 - 1110.

Koslofe, D., Scott, R. F. and Scranton, J. (1980): Finite element simulation of Wilmington oil field subsidence. Nonelinear modelling. Tectonophysics, v 70, pp 159 - 183.

Kovanic, L. (Jr.), Kristakova, Z., Lescinsky, I. and Sedlak, V. (1988): Mapping of leached spaces in a salt deposit and determination of the undermined ground deformations. Proc. CONFERENCE 1988, Fredericton, New Burnswick, Canada, pp 279 - 288.

Kowalezyk, Z.(1966): Effects of mining exploitation on the ground and structures in heavily industrialised and populated area. CIMM Bull. Trans., LXIX, n10, pp 1201 - 1207.

Krishna, R., Kumar, L. A., Raju, M. V. B. and Chattopadhaya, T.(1989): The current knowledge of ground movement due to mining. Land Subsidence, Oxford & IBH Pub., New Delhi, pp 121 - 132.

Kruse, G.A.M. (1995) : Subsidence at the transition from massif to basins in the southwestern, Netherlands. Land Subsidence, IAHS Publication, No.234, pp 59-68.

Kuecher, G.J. (1995) : The dominant processes responsible for subsidence of coastal wetlands in South Louisiana. Land Subsidence, IAHS

Publication, No.234, pp 69-82.

Kuhn, P.(1988): Monitoring of tilt and inclination of engineering structures. Proc. CONFERENCE 1988, Fredericton, New Burnswick, Canada, pp 289 - 297.

Kumar, B. (1991): Mechanics of Ground Subsidence in Indian Coalfields. Ph.D. Thesis, Indian School of Mines, Dhanbad.

Kumar, B., Saxena, N.C. and Singh, B.(1983): A new hypothesis for subsidence prediction. J. Min. Met. & Fuels, pp 459 - 465.

Kumar, B., Saxena, N. C. and Singh, B. (1983): Fragmentation of superincumbent strata and subsidence movements due to extraction of coal seams underground. Proc. Workshop on Rock Fragmentation, B.H.U., Varanasi.

Kumar, R. (1971): Investigations Into Surface Subsidence Due to Underground Mining. Ph. D. Thesis, B.H.U.,Varanasi.

Kumar, R., Singh, B. and Sinha, K. N. (1973): Subsidence investigations in Indian mines. Proc. Symp. Subsidence in Mines, Aust. Inst. Min. Met., pp 6/1 - 6/7.

Kuwahara,T. et.al.(1976): Analysis of land subsidence in the Nobi plain. Land Subsidence, IAHS Publication No 121, pp 55 - 64.

Lambeck, K. and Johnston, P. (1995) : Land subsidence and sea- level change : Contributions from the melting of the last great ice sheets and the isostatic adjustment of the Earth. Land Subsidence, IAHS Publication, No.234, pp.3-18.

LaScola, J. C. and Trevits, M. A.(1988): Longwall mine subsidence surveying : An engineering technology comparison. Proc CONFERENCE 1988. Fredericton, New Burnswick, Canada, pp 298 - 308.

Lacey, W.D. and Swain, N.T.(1973): Design for mining subsidence. Archit. (Build.) J., v 126, p 557.

Lakington, D. W. and Robinson, B. (1973): Articulated service reservoirs in mining subsidence areas. Inst. Wat. Engr. J., v 27, n 4, p 197.

Land Subsidence(1977), IAHS Publication No. 121, 669 p.

Land Subsidence(1984), IAHS Publication No.151, 939 p.

Land Subsidence(1989), Oxford & IBH Pub., New Delhi, 673 p.

Land Subsidence(1991), IAHS Publication No. 200, 690 p.

Landes, K. K. and Piper, T.B. (1972): Effect Upon the Environment of Basin Cavity Subsidence at Grosstle, Michign - 1972. Solution Mining Research Institute, 812 Musiel Street, Woodstock, v Ill, 60098., 52 p.

Lane, W.T. and Roberts, J.H. (1929): The principles of subsidence and the law of support. Alfred A. Knoff., 319 p.

Larson, M. K. (1984): Potential for subsidence fissuring in the Phoenix, Arizona, USA area. Land Subsidence, IAHS Publication No 151, pp 291 - 300.

Lautsch, H. (1969): Principles of theoritical prediction of ground movements - comparison with practical results and the task of mining surveyor in the application of these predictions. Proc. 1st. Can. Symp. on Mine Surveying and Rock Deformation Measurements, pp 293 - 324.

Leal, J., Murria, J. and Pedroza, M. (1995) : Recent developments in subsidence monitoring and prediction in the Costa Oriental Oilfields, Venezuela. Land Subsidence, IAHS Publication, No.234, pp.179-186.

Leake, S.A. (1991): Simulation of vertical compaction in models of regional ground water flow. Land Subsidence, IAHS Publication No 200, pp 565 - 574..

Lee, A. J. (1965): The Effect of Faulting on Subsidence. M. Sc. Thesis, Univ. of Nottingham, UK, 136 p.

Lee, A.J. (1966): The effect of faulting on mining subsidence. The Min. Engr., v 55, pp 735 - 745.

Lee, C. K., Fallou, S. N. and Mei, C. C. (1991): Subsidence due to intensive pumping from a layered soil. Land Subsidence, IAHS Publication No 200, pp 677 - 690.

Lee, F. T. and Abel, J. F. (1982): Subsidence form Underground Mining : Environmental Analysis and Planning Considerations. US geol. Survey Circular, 876.

Lee, K. L. and Shen, G.K. (1969): Horizontal movements related to subsidence. J. Soil Mech. Found. Div. Proc. ASCE, n 95 (SM1), pp 139 - 166.

Leeds University(1957): Proceedings of the European Congress on Ground Movement, 218 p.

Legal, Socio-Economic and Environmental Significance of Land Subsidence in the United States.(1981), Amr. Soc. Civil Engrs. J. Izzig. Div., v 107, n IR2, pp 113 - 185.

Legget, R. F. (1962): Geology and Engineering. Mc Gra-Hill Book Company, NewYork, 884 p.

Lenaz, R. and Taviani, M. (1984): Differentiated subsidence of upper Adriatic area facing Vanice, Land Subsidence, IAHS Publication No 151, pp 359 - 366.

Lenge, A. (1957): Inspection and correction of damage to railways caused by mining in the Saar. Proc. Eur. Cong. Gr. Movement, pp 106 - 114.

Lenge, A. (1958): Inspection and correction of damage to railways caused by mining in Saar. Proc. 2nd. St. Cont. Cong., Leipzig.

Levanon, D. and Levin, I. (1984): Preservation of peat soil by cultivation of prernnial herbage crops. Land Subsidence, IAHS Publication No 151, pp 529 - 537.

Levin, I. and Shoham, D. (1983): Ground surface elevation loss in the reclaimed Hula swamp in Isreal during the period 1958- 1980. Proc.

2nd. Int. Symp. on Peat in Agriculture and Hoticulture, Bet Dagan, Israel.

Liao, J. S., Pan, K. L. and Haimson, B. C.(1991): The monitoring and investigation of ground subsidence in southwestern Taiwan. Land Subsidence, IAHS Publication No 200, pp 81 - 97.

Lin, G. M. (1989): Land subsidence analysis with Monte-Carlo simulation method. Land Subsidence, Oxford & IBH Pub., New Delhi. pp 147 - 156.

Lin, P. M., Hsiung, S. M. and Peng, S. S.(1987): Investigation of subsidence over AML. A case study. Proc. 6th. Int. Conf. Gr. Cont. in Mines, Morgantown, WV, USA, pp 249 - 257.

Lin, S., Whittaker, B. N. and Reddish, D.J.(1992): Application of asymmetrical influence functions for subsidence perdiction of gently inclined seam extractions. Int. J. Rock Mech. & Min. Sci. & Geomech. Abstr., v 29, n 5, pp 479 - 490.

Lisowski, A.(1956): Rock displacements according to the theory of elastcity. Archiwum gornictwa, v 1, pp 76 - 94.

Liszkowski, J.(1991):Engineering and environmental impacts caused by land subsidence due to sub-surface extraction of solid raw materials from Poland. Land Subsidence, IAHS Publication No 200, pp 369 - 378.

Littlejohn, G. S.(1979): Surface stability in areas underlain by old coal workings. Ground Engineering, v 12, n 2, pp 22-30.

Litwiniszyn, J.(1957):The theories and model research of movements of ground mass. Proc. Europ. Cong. Gr. Movement, pp 202 - 209.

Litwiniszyn, J.(1959): Thiorie des movements de terrains et recherche sur modeles. Ind. Min., v 41, n 5, pp 433 - 446.

Litwiniszyn, J.(1980):The influence of ground surface deformations caused by underground mining on the state of stress in embakments. Proc. 2nd. Int. Symp. at Univ. of Wales, Cardiff, pp 92 - 98.

Lloyd, W. D.(1918-19): The effect of coal mining on the overlying rocks and on the surface. Trans. Inst. Mining Engrs., v 57, pp 74 - 93.

Lorenz, G.K., Beusekom, W.van, Groenewoud, W. and Hofman, M. (1995) : Geodetic determination of land subsidence in The Netherlands. Land Subsidence, IAHS Publication, No.234, pp.187-196

Lofgren, B. E.(1961): Measurement of Compaction of Aquifer System in Areas of Land Subsidence. US Geological Survey Professional Paper 424-B., pp B 49 - B 52.

Lofgren, B. E. (1969): Land subsidence due to the application of water. Reviews in Engineering Geology II : Geological Society of America, v 2, pp 271 - 303. Boulder Co.

Lofgren, B. E. (1969): Field measurement of aquifer - system compaction, San Joaquin Valley, California, USA. Land Subsidence, IAHS Publication. No 88, pp 272 - 284.

Lofgren, B. E. (1978): Change in aquifer system properties with ground water depletion. Proc. Int. Conf. on Evaluation and Prediction of Subsidence, Pansaccola Beach, Florida, pp 26 - 46.

Lofgren, B.E. and Ireland, R.L. (1973): Preliminary Investigation of Land Subsidence in the Sacramento Valley, California, US Geological Survey Open-File Report, 40 p.

Louis, H.(1922):A contribution to the theory of subsidence. Trans. Inst. Mining Engrs., v 64, 257 p.

MaCaul, C., Morgan, J. M. and Boden, J.B. (1976): Measurement of Ground Movement Due to Extraction of a Shallow Tunnel in Lower Chalk. T.R.R.L. Supplementary Report 199, US Dept. of Environment.

Macccourt, L., Madden, B. J. and Suchmann, E. H. R. (1986): Case studies of surface subsidence over collapsed bord and pillar workings in South Africa. Proc. SANGORM, 1986 Symp., pp 25 - 32.

Madhav, M. R. and Basak, P. (1977): Ground subsidence due to nonlinear flow through deformable porous media. J. Hydrology, v 34, pp 21 - 33.

Mahar, J. W. and Marion, G. G. (1982): Building responce and mitigation measures for building damage in Illinois. Proc. Workshop on Surface Subsidence due to Underground mining. Morgantown: West Virginia Univ., pp 238 - 252.

Maiza. E. R., Thomas, E. and Greenwald, H. F. (1941): Studies of Roof Movement in Coal Mines. Studies of Subsidence of a Highways Caused by Mining Coal Beneath. USBM Report 3562, 11 p.

Malhotra, K.K. and Rai, P.N. (1989): Land subsidence due to fires and measures taken to combat the effects on structures - a case study. Land Subsidence, Oxford & IBH Pub., New Delhi. pp 421 - 430.

Marco, G. and Paolo, R. (1995) : Land subsidence in Po river valley. Land Subsidence, IAHS Publication, No.234, pp 149-158.

Marabini, F. and Ricceri, G. (1984): Highlights of oil , gas and mineral extraction succession. Land Subsidence, IAHS Publication No 151, pp 937 - 938.

Maraffi, C., Mozzi, G., Posocco, F., Rusconi, A., Salzano, E., Santangelo, R., Sortino, L. and Zanetti, L. (1984): Venice today. Land Subsidence, IAHS Publication No 151, pp 329 - 336..

Marino, G. G. (1986): Mine subsidence damage from room and pillar mining in Illinois. Int. J. Min. & Geological Engg., v 4, pp 129 - 150.

Marino, G.G. (1987): The initial collapse of the overburden over longwall panels. Proc. 6th. Int. Conf. Gr. Cont. in Mining, Morgantown, WV, USA, p 183.

Marino, G. G. and Cording, E. J. (1985): Geotechnical aspects of subsidence over room and pillar mines in Illinois. Proc. 4th. Conf. Gr. Cont. in Mining, Morgantown, WV, USA, pp 240 - 248.

Marino, G.G. and Mahar, J.W.(1985): Subsidence Damaged Homes Over Room and Pillar Mines in Illinois. Report for Illinois Abandoned Mined Lands Reclamation Council, Springfield, Illinois, 450 p.

Marino, G. G. et. al. (1986): Ground Subsidence and Structural Damage Over an Abandoned Room and Pillar Coal Mine at Hegeler, IL. Bureau of Mines Information Circular 9072.

Mark, C. and Bieniawski, Z. T. (1987): A new method for sizing longwall pillars based on field measurements. Proc. 6th. Int. Gr. Cont. in Mining, Morgantown, WV, USA, pp 157 - 171.

Marr, J. E. (1949): The estimation of mining subsidence. Coll. Guard,v 198, n 5116., pp 345 - 352.

Marr, J. E. (1959): A new approach to the estimation of mining subsidence. Trans. Inst. Min. Engrs., v 118, pt 11, p 692-706.

Marr, J.E.(1961): Subsidence observations in the South Lancashire coalfield. Sheffield Univ. Mining Magazine, pp 24 - 35.

Marr, J. E. (1965): The effect on surface property by a modified mining method. Chart. Surv., v 97.

Marr, J. E.(1975): The application of the zone area system to the prediction of mining subsidence. The Mining Engineer, v 176, pp 53 - 60.

Marrino, G. G. and Mahon, J. W.(1985): House damage criteria over Illinois room and pillar coal mines. SME of AMIE Transactions, v 278, pp 1818 - 1825.

Marsily, G. de., Ledoux, E., Levassor, A., et.al. (1978): Modelling of large multi-layered aquifer system : theory and applications. J. of Hydro., v 36, pp 1 - 33.

Martin, J. C. and Serdengecti, S.(1984): Subsidence over oil and gas fields. Man Induced Land Subsidence. Rev. Eng. Geology VI, The Geological Soc. of America, pp 23 - 34.

Martos, F.(1958): Uber eine Naherungsgleichung der Senkungs mulde und dezen Zeitfaktor. Proc. Int. Bureau of Rock Mech., Berlin, pp 102 - 109.

Martos, F.(1958): Concerning an approximate equation of the subsidence trough and its factor. Proc. Int. St. Cont. Cong., Leipzig, pp 191 - 205.

Matheson, G. M. and Pearson, M. L. (1985): Chimney Subsidence Development in the Colorado Spring Coalfield, Colorado. SME - AIME Fall Meeting, Albuquerque, New Mexico, Preprint No.85-329.

Mathur, S. P. and Prasad, N.(1981): Coal extraction under surface features - harmonious mining. Mine Subsidence, Parijat Mudr. Pub., Dhanbad, pp 108 - 125

Matsuoka, I. (1984): Land subsidence of an inland basin, Northeastern Japan - The Yamagata Basin. Land Subsidence, IAHS Publication No 151, pp 373 - 380.

Maurer, W., RoBmeier, F. and Schriadelbach, K.(1988): Determination of periodic displacements of buildings and machines with the aid of a laser interferometer. Proc. CONFERENCE 1988, Fredericton, New Burnswick, Canada, pp 339 - 347.

Maurizio, B., Donatella, D., Gabriele, F. and Fabio, R.(1989): GPS monitoring of Ancova landslide. Land Subsidence, Oxford & IBH Pub., New Delhi. pp 353 - 364.

Maury, V., Sauzary, J. M. and Fourmanitraux, D.(1987): Geomechanical approach to oil and gas production - main problems, first results. Proc. ISRM 6th. Int. Cong. Rock Mech., v 1, pp 181 - 189.

Mazurkiewiez, M.(1992): Thick coal mining with roof caving and backfilling. Thick Seam Mining, Oxford & IBH Pub., New Delhi, India. pp 287 - 298.

Mc Callum,T.(1952):The maintenance of roads in a mining area. J. Inst. Highway Engrs., v 2, p 6.

Mc Callum, T. (1943-44): Mineral subsidence and local authority services. Proc. Inst. Munic. Engrs., v 70, p 411.

McNulty, E.G.(1991): Sand drain induced subsidence of a peat. Land Subsidence, IAHS Publication No 200. pp 333 - 342.

MeNaab, K. E.(1987): Three Dimensional Numerical Modelling of Surface Subsidence Induced by Underground Mining. CSIRO Tech. Report No. 146. Div. of Geomech., CSIRD, Viet., Australia, 20p.

Meade, R.H.(1967): Petrology of Sediments Underlying Areas of Land Subsidence in Central California. US Geological Survey Professional Paper, 497-C, 83 p.

Meade, R. H.(1968): Compaction of Sediments Underlying Areas of Land Subsidence in Central California. US Geological Survey Professional Paper., 497-D, 39 p.

Mendoza, H. and Murria, J.(1989): Ground subsidence modelling in Western Venezuela. Land Subsidence, Oxford & IBH Pub., New Delhi. pp 133 - 146.

Metrusty, J.W.(1984): Pressure effects in coal mines. Coll. Engg., v 25, pp 134 - 137.

Meyer, W.R. and Carr, J.E. (1979): A Digital Model for Simulation of Groundwater Hydrology in the Houston Area, Texas. Texas Dept. of Water Resources Report LP - 103, 131 p.

Miao, J. F. and Wu, L. G. (1991): Mathematical modelling of land subsidence due to a multi-aquifer system with visco-elastic properties. Land Subsidence, IAHS Publication No 200, pp 593 - 602.

MiKula, P. A. and Holt, G. E.(1983):Prediction of mine subsidence in eastern Australia by mathematical modelling. Proc. 5th. Int. Cong. Rock Mech., Melbourne, Australia, pp E 119-E 125.

Milev, G. K.(1988): Generalized dynamic models with stress and deformations. Proc. CONFERENCE 1988, Fredericton, New Burnswick, Canada, pp 348 - 358.

Miller, H. D. S. and Stoakes, M. A.(1985): Subsidence associated with Lonisiana Salt domes. Proc. CLM Conf. on Subsidence in Soft Rock Mining, Saskatoon, Saskatchewan, 20 p.

Mine Subsidence(1986): SME Fall Meeting, St.Louis, MO, USA, Sept, SME, Littleton, CO, USA, 143 p.

Mining 88 (1988): Proceedings of the Second International Conference on Constructions in Areas of Abandoned Mine Workings, Edinburgh.

Ministry of Works(UK) (1951): Mining Subsidence, Effects on Small Houses. Nat. Build. Studies Spec. Rep., n 12, H.M.S.O. London.

Misich, I.(1993): An integrated approach to subsidence modelling and prediction in the Collie Basin (Western Australia). Proc. 12th. Int. Conf. Gr. Cont. in Min., Morgantown, WV, USA, pp 404 - 411.

Missavage, R. J. et.al.(1986): Subsidence prediction in shallow room and pillar mines. Int. J. Min. Engg., v 4, pp 39 - 46.

Monem, Abd El, and Mohamed, S. (1995) : Horizontal and vertical crustal deformation studies along the Seiyal fault, northwest of Aswan Lake, Egypt. Land Subsidence, IAHS Publication, No.234, pp103-112.

Moebs, N. N. and Barton, T. M.(1985): Short-term effects of longwall mining on shallow water sources. Mine Subsidence Control. Proc. of Bureau of Mines Technology Transfer Seminar, Pittsberg, Pa, Bureau of Mines IC 9042, pp 13 - 24.

Molesky, P.J.(1988): Surface Ground Movements Over Longwall Mining in the Pitsburg Seam. M.S. Thesis, Dept. of Min. Engg., West Virginia Univ., WV, USA.

Montz, H.W.(1930): Subsidence from anthracite mining. Trans. AIMME Coal Div., pp 98 - 134.

Moore, H. L.(1980): Karst problems along Tennessce highways, an overview. Highway Geology Reprint, 31st. Annual Symp., Austin, Texas, 68 p.

Moore, R.C. and Nawrocki, M.A.(1980): Effects of Subsidence form Thick Seam Coal Mining on Hydrology. US Bureau of Mines, NTIS No. PB80-21980.

Morales, R., Muzillo-Fernaadez, Y. and Hernandez-Rubio, A. (1991): Subsidence of the former Texcoco lake. Land Subsudence, IAHS Publication No 200. pp 35 - 44.

Morrison, R. G. K.(1976): A Philosophy of Ground Control. Dept. of Min. & Met. Engg., McGill Univ. Canada, 182 p.

Morrison, W. C.(1991): Grouting in deep flooded mines. Land Subsidence, IAHS Publication No 200. pp 503 - 514.

Morse, C. F. R.(1967): Some mining problems (M6 Motorway). The Chartered Surveryor, Nov. '67, p 236.

Moscal, S. J.(1988): Subsidence Movement Due to Longwall Mining. M.S.

Thesis, Dept. Min. Engg., West Virginia Univ. Morgantown, WV, USA.

Mozumdar, T. K.(1985): Subsidence in Indian Coal Mines. M. Tech. Thesis, ISM, Dhanbad, India.

Mozumdar, T. K. and Mozumdar, B. K.(1989): Subsidence studies in coal mines at shallow depth. Land Subsidence, Oxford & IBH Pub., New Delhi. pp 497 - 516.

Mraz, D.Z.(1985): Proposed Subsidence Model for Saskatchewan Potash Mine. Paper presented to Semnar of RMSC Sub committee on Soft Rocks, 9 p.

Mukherjee(Basu), R.(1989): Environmental, socio-economic and legal impacts of mine subsidence in a part of Ranigunj coalfield, West Bengal. Land Subsidence, Oxford & IBH Pub., New Delhi, pp 555 - 564.

Mukherjee, S.K. (1989): A few cases of land subsidence in Ranigunj coalfield over old unapproachable workings decleared unsafe - a critical review. Land Subsidence, Oxford & IBH Pub., New Delhi, pp 193 - 206.

Muller,W.(1973): Die Berechnung der Bewegungen and Spannungen des Gebirges vom Abbau bis zur Tapesoberflache nach der Method der endlichen Elemente. Diss. Berlin.

Munjeri, D.(1987): Pervention of subsidence using stowing methods. Colliery Guardian, v 235, n 7, pp 244,246 - 247,250.

Munson, D.E. and Eichfeld, W.F.(1980): European Empirical Methods Applied to Subsidence in the Coalfields. Sandia National Laboratories, Report SAND 80 - 1920, Albuqueique, New Mexico.

Munson, D. E. and Eichfeld, W. F.(1980a): Evaluation of European Empirical Method for Subsidence in US Coalfields. SAND 80 - 0537, Sandia National Laboratories Albuquerque, New Mexico.

Munson, D.E. and Southerland, H.J.(1982): Empirical and room and pillar analytic approaches to subsidence prediction. Ground Control in Mining, SME, pp 139 - 149.

Murria, J.(1991): Subsidence due to oil production in Western Venezuela : Engineering problems and solutions. Land Subsidence, IAHS Publication No 200. pp 129-141.

Murria, J. and Abi Saab,J.(1988): Engineering and construction in areas subjected to subsidence due to oil production. Proc. CONFERENCE 1988, Fredericton, New Burnswick, Canada, pp 367-373.

N.C.B. Production Department, (1952): Investigation of Mining Subsidence Phenomenon. Information Bulletin 52/78.

N.C.B. Production Dept.(1961): Partial Extraction as a Means of Reducing Subsidence Damage. Information Bulletin 61/78.

NCB,(1963): Principles of Subsidence Engineering. NCB Publication Dept. Information Bulletin, 61/240, 27 p.

NCB(1966):Subsidence Engineering Handbook, NCB Plublication Dept., 130 p.

NCB, (1968):Working Under the Sea. NCB Publication Dept.Instruction, PI/1968/8, 4 p.

NCB, (1975): Subsidence Engineers' Handbook. 2nd. ed. London: UK National Coal Board Production Dept.

NCB,(1982): Mine Subsidence Report. NCB, Northeast Leicestershire Prospect, Bestwood, Nottingham, U.K., 37 p.

Nakao, T. and Kamata,A. (1991): Counter measures for and monitoring of land subsidence in the northwestern part of the Kanto Plain, Japan. Land Subsidence, IAHS publication No 200, pp 525 - 532.

Narasimhan, T.N. and Witherspoon, P.(1977): Numerical model for saturated - unsaturated flow in deformable porous media, 1 Theory. Water Resource Research, v 13, n 3, pp 657 - 664.

Narayan, S., Raju, N. M. and Singh, B.(1989): Microseismic method applied to failure warning in Indian coal mine. Land Subsidence, Oxford & IBH Pub., New Delhi. pp 589 - 597

National Contribution by 12th Land Subsidence Committee Members. 1989: Japan (Akagi, T.), The Nitherlands (F.B.J.Barends), United Kingdom (Paul Cater), Swden (S.Hansbo), Venezuela (J. Muzzia), Italy (G. Ricceri), India (N.C. Saxena), and South Africa (F.Wagner). Proc. 12th Int. Conf.on Soil Mech. and Foundation Engg., Rio De Janeiro, A.A.Balkema, pp 3205-3220.

Neal, J.T.(1991): Prediction of subsidence resulting from creep closure of solution mined caverns in salt domes. Land Subsidence, IAHS Publication No 200. pp 225 - 234.

Neighbors, R. J. (1981): Subsidence in Harris and Galveston countries, Texas. ASCE J. Irrgi. Drain.Div., v 107,n IR2, pp 162 - 174.

Neighbors, R.J. and Thompson, R.E.(1984): Subsidence in the Houstan-Galveston area of Texas. Land Subsidence, IAHS Publication No 151, pp 785 - 794.

Neuhaus, W. (1975): Die Fernwirkung eines Abbauekmentes bei elastoplastischem Gerbirge. Gluckauf- Forsch, v 36, pp 199 - 205.

Neuhaus, W. (1976) : Die Berchnungsgrand lagen der bergschaderkundlichen Einwirkungsuetre and iher Moglichkeiten and Grenzen. Mitt. Markscheidew, v 83, pp 41 - 60.

Neuman, S. P., Preller, C. and Narasimhan, T. N. (1982): Adative explicit - implicit quasi three-dimensional finite element model of flow and subsidence in multiaquifer system. Water Resource Research, v 18, n 5, pp 1551 - 1561.

Newton, J. G.(1976): Early Detection and Correction of Sinkhole Problem in Alabama, with a Preliminary Evaluation of Remote Sensing Application. Alabama Highways Dept Bureau Research and Development, Research Report No. HPR - 76, 83 p.

Newton, J. G.(1984): Natural and induced sinkhole development in the Eastern United States. Land Subsidence, IAHS Publication No 151, pp 549 - 564.

Newton, J. G. (1984): Sink-holes resulting from ground water withdrawals in carbonate terranes - an overview. Man Induced Land Subsidence, Rev. Eng. Geology VI, The Geological Soc. of America, pp 195 - 202.

Niederhofer, G. (1962): A new method of precalculating surface subsidences, particularly over steep workings. Mitteilungen aus dem Markscheidewęsen, v 69 No.2. pp 33 - 75.

Niemczyk, O. (1949): Bergschadenkunde. Gluckauf, Essen.

Niemczyk, O. (1949): Study of mining damages, Verlag Gluckauf, Essen, W. Germany, 291 p.

Nieto, A.S.(1979): Evaluation of damage potential to earth dam by sub-surface coal mining at Rend Lake, Illinois. Proc. 10th Ohio River Valley Soils Seminar on Geotechnics in Mining, Lexington, Kentucky, pp 9 - 18.

Nishida, T., Esaki, T., Akoki, K. and Kameda, N.(1983): Evaluation and Prediction of Subsidence in Old Working Areas and Preventive Measures from Mining Damage for Newly Constructing Structures. Report of Research, Institute of Industrial Sci., Kyushu University, No. 74.

Nishida, T., Esaki,T., Aoki, K and Kameda, N.(1984): Evaluation and prediction of subsidence in old working areas and practical preventive measures against mining damage to new structures. Land Subsidence, IAHS publication No151, pp 717 - 726.

Nisida, T., Esaki, T., Aoki, K. and Kameda, N.(1984a): Land subsidence in natural gas fields. Land Subsidence, IAHS Publication No 151, pp 699 - 705.

Norasimham, T.N. and Goyal, K.P.(1984): Subsidence due to geothermal fluid withdrawal. Man Induced Land Subsidence Rev. Eng. Geology VI, The Geological Soc. of America, pp 35 - 66.

North, F. J.(1952): Some geological aspects of subsidence not due to mining. South Wales Institute of Engineers, v 67, pp 127- 153.

Nutalaya, P., Chandra, S. and Balasubramaniam, A.S.(1984): Subsidence of Bangkok city due to deep well pumping and its control to artificial recharge. Land Subsidence, IAHS publication No 151, pp 727 - 745.

Q'Connor, K. M. and Dowdling, C. H. (1988): Monitoring and simulation of mining induced subsidence. Proc. CONFERENCE 1988. Fredericton, New Burnswick, Canada, pp 390 - 399.

O'Donahue, T. A. (1929-30): Subsidence caused by coal mining. Trans. I. Min. E., v 78, pp 91 - 110.

O'Reilly, M.P. and New, B.M.(1982): Settlements above tunnels in the United Kingdom, their magnitude and prediction. Proc. Tunnelling '82, Brighton, Discussion Trans. I. M. M. 1983, A35 - A48, pp 171 - 181.

O'Rourke, J. E., Mabry, R. M., Ranson, B. B. and O'Connor, K.(1977): Subsidence Monitoring for Undermined Areas. Final Report Submitted to US DOE under Contrat No.ET-76-C-01-9123, 340 p.

O'Rourke, T. D. and Turner, S. M. (1979): Longwall subsidence patterns, a review of observed movements, controlling parameters, and empirical relationships. Geotechnical Engg., Report 79(6):82 p.

O'Rourke, T. D. and Turner, S. M.(1981): Empirical methods for estimating subsidence in US coalfields. Proc. 22nd. US Symp. Rock Mech., Massachusetts Institute of Technology, Cambridge, pp 322 - 327.

Obert, L. and Duvall, W. I. (1967): Rock Mechanics and Design of Structures in Rock. Wiley, NewYork, pp 544 - 572

Ogden, H. and Orchard, R.J. (1960): Ground movements in North Staffordshire. Trans. Inst. Min. Engrs., v 119, IV, pp 259 - 272.

Oglobin, D.M. (1960): Basic laws governing the process of rock movement in working several shallow - dip seams. Freiberger Forschungshefte, A 185, pp 63 - 76.

Ohshima, T., Ueshita, K. and Daito, K.(1989): Land subsidence and groundwater condition in Nagoya. Land Subsidence, Oxford & IBH Pub., New Delhi. pp 283 - 294.

Once, G.(1985): Application of finite element analysis to the ground control problems associated with mining of steeply inclined chromite ore body in Turkey. Proc. 5th Int. Conf. Num. Mech., Geomech., Nogoya., pp 1589 - 1594.

Orchard, R.J.(1964): Surface subsidence resulting from alternative treatment of colliery goaf. Colliery Engg.

Orchard, R. J.(1957a): Prediction of the magnitude of surface movements. Proc. Eurp. Cong. Gr. Movement, pp 39 - 46.

Orchard, R.J.(1957b): The effects of mining subsidence upon public health engineering work. J. Inst. Publ. Health Engrs., v 56, p 188.

Orchard, R. J.(1956-57): Surface effects of mining - the main factors. Trans. I. Min. E., v 116, pp 941 - 955.

Orchard, R.J.(1954): Recent developments in predicting the amplitude of mining subsidence. J. R. Instn. Chart. Surv., v 33, 864 p.

Orchard, R.J.(1957c): Prediction of the magnitude of surface movements. Colliery Engg., v 34, 455 p.

Orchard, R.J.(1969): The control of ground movements in under sea working. Min. Engr., v 128, p 259.

Orchard, R. J. and Allen, W.S.(1965): Ground curvature due to coal mining. Chart Surv., v 97, n 11, pp 622 - 631.

Orchard, R.J. and Allen, W.S.(1970): Longwall partial extraction systems. Mining Engineer., v 129, pp 529 - 535.

Padmanabhan, G.(1981): Coal extraction under railway aquired land in India - associated problems. Mine Subsidence, Parijat Mudra. Pub.,

Dhanbad. pp 47 - 50

Pandey, B. N., Prasad, S. N., Mukherjee, K. P., Tiwary, S., et.al.(1992): Experince of blasting gallery method of thick seam mining in Jharia and Ranigunj coalfield. Thick Seam Mining, Oxford & IBH Pub., New Delhi, pp 409 - 422.

Panek, L.A.(1966): Methods and equipment for measuring subsidence. 3rd. Symp. on Salt, Cleveland: Northern Ohio Geological Society, Inc., v 2.

Papo, H. B. and Perelmuter, A. (1988): The datum issue in deformation analysis. Proc. CONFERENCE 1988, Fredericton, New Burnswick, Canada, pp 400 - 409.

Pariseau, W.G. and Dahl, H.D.(1968): Mine subsidence and model analysis. Trans. AIME, v 241, pp 488 - 494.

Park, D. W. and Nadim, F. A. (1987): Numerical modelling of progressive subsidence. Proc. 15th. Annual Work. Generic Mineral Tech. Centre Mine Design and Ground Control, Virginia Polytechnic Institute and State Univ., pp 111 - 118.

Park. D.W.(1988): Two case histories of subsidence in the Wazzior coalfield. Min. Engr., pp 185 - 191.

Pasamehmetoglu, A.G.(1972): An Investigation Into Time-Dependent Aspects of Mining Subsidence. Ph.D. Thesis, Univ. of Nottingham, UK., 190 p.

Peaker, K. R., Gore, I. and Ahamd, S. A. (1991): Controlled subsidence during pile driving. Land Subsidence, IAHS Publication No 200, pp 359 - 368.

Peck, R.B.(1969): Deep excavation and tunelling in soft ground. Proc. 7th. Int. Conf. Soil Mech. and Found. Engg. Mexico City.

Peng, S. S.(1980): Roof Fall in Coal Mines. Dept. of Min. Engg., West Virginia Univ., TR80-11, 56 p.

Peng, S.S.(1986): Coal Mine Ground Control. 2nd. Ed. John Wiley & Sons, Inc. New York, pp 295 - 313.

Peng, S.S.(1990): Comments on surface subsidence prediction. Min. Sci. & Tech., v 11, pp 207 - 211.

Peng, S.S. and Chandra, U.(1980): Getting the most from multiple- seam reserves. Coal Mining and Processing, pp 78 - 84.

Peng, S. S. and Cheng, S. L. (1980): Prediction of subsidence profile due to underground coal mining. Coll. Min. & Energy Res. WV Univ. Tech Report., Tr-80-5, 46 p.

Peng, S. S. and Geng, D. Y. (1982): Methods of predicting the subsidence factor, angle of draw, and angle of critical deformation. Proc. of the state of art of Gr. Cont. in Longwall Min. and Min. Subsidence, pp 211 - 215.

Peng, S. S. and Geng, D. Y.(1983): The Appalachian Field; general characterisatics of surface subsidence and monitoring methods. Surface Min. & Envr. Moni. Recla. Handbook, Elsevier, pp 627 - 646.

Peng, S.S. and Luo,Y.(1988): Monitoring and prediction of surface dynamic subsidence due to longwall coal mining. Proc. CONFERENCE 1988, Fredericton, New Burnswick, Canada, pp 420 - 429.

Peng, S.S. and Geng, D.Y.(1982): Surface subsidence, overburden behaviour and structural damages due to longwall mining - two case studies. Proc. 2nd Int. Cong. Stability in Underground Mining, Univ. of Kantucky, pp 497 - 523.

Perz, F.(1959): A new method of expressing the mathematical relationships governing the formation of subsidence troughs above mine workings. Proc. Europ. Cong. Gr. Movement, Leeds, pp 21 - 26.

Perz, F.(1948): The influence of time on ground movements above workings. Mitteilungen aus dem Markscheidewesen, v 55, pp 92 - 117.

Perz, W.(1957): Observation of subsidence over underground working in the Austrain Tertiary Coal Basins. Proc. European Cong. Gr. Movement, Leeds., pp 92 - 98.

Pflagging, K.(1989): 10 years experience with aerophotogrammetric measurements using GPS for recording ground movements. Land Subsidence, Oxford & IBH Publ., New Delhi. pp 35 - 36.

Pflagging, K.(1984): Quantifying the effects of continuous underground mining. Ground Movements & Structures, Int. Conf. at Univ. Of Wales, Cardiff, pp 354 - 376.

Phiancharoen, C. and Buapeng, S.(1973): Hydrogeology of the Lower Chao Phraya Delta. Ground Water Bull. No.4, Dept. of Mineral Sources, Bangkok.

Phien-wej, N., Nutalaya, P. and Sophonsakultat, W.(1989): Current land subsidence of Bangkok and contemplated remedial measures. Land Subsidence, Oxford & IBH Publ., New Delhi. pp 311 - 322.

Phillips, D.W. and Henshaw, H.(1942): Underground and surface strata movements. Trans. Inst. Min. Surv., v 21, pt.1.

Pieri, L. and Russo, P. (1984): The survey of soil vertical movements in the region of Bologna. Land Subsidence, IAHS Publication No 151, pp 235 - 246..

Piggot, R. J. and Eynon, P.(1977): Ground movements arising from the presence of shallow abandoned mine workings. Large Ground Movements and Structures, Pentech.pp 749 - 780.

Piyasena, W.M.(1982): Rehabilitation of a Depleted Aquifer System Through Artificial Recharge - An Application to Bangkok. A.I.T.D. Eng. Dissertation No. GT-82-1,Bangkok, 327 p.

Poland, J. F.(1961): The Coefficient of Storage in a Region of Major Subsidence Caused by Compaction of An Aquifer System. In Geological Survey Research. US Geological Survey Professional

Paper 424-B, pp B 52 - B 54.

Poland, J.F.(1972): Subsidence and its control. Underground Waste Management and Environmental Implications, Memoir No. 18, Houstan, Texas, Aus. Assoc. Pet. Geologists, pp 50 - 71.

Poland, J.F.(1981): Subsidence in United States due to groundwater withdrawal. ASCE J. Irrig. Drain. Div., v 107, n IR2, pp 115 - 135.

Poland, J. F., Lofgren, B.E., Ireland, R.L. and Pugh, R.G. (1975): Land subsidence in San Joaquin Valley, California, as of 1972. US Geological Survey Professional Paper 437-H, 78 p.

Poland, J.F., and Yamomoto, S.(1984): Field Measurement of Deformation. Guidebook to Studies of Land Subsidence Due to Ground-water Withdrawal, UNESCO, Peris, France, pp 17 - 35.

Pomeray, J. S. (1986): USGS investigation of landslide susceptibility. J. of Gr. Failure, v 3, 24 p.

Potok, A.J.(1991): A study of the relationship between subsidence and flooding. Land Subsidence, IAHS Publication No 200, pp 389 - 386.

Pottgens, J. J. E.(1978): Ground movements by coal mining in the Netherlands. Proc. Conf. Evaluation and Prediction of Subsidence, A.S.C.E. Pensacola Beach, Florida, pp 267 - 282.

Pottgens, J. J. E. (1984): Ground movements caused by mining activities in the Netherlands. Land Subsidence, IAHS Publication No151, pp 651 - 660.

Pottgens, J. J. E. and Brouwer, F. J. J.(1991): Land subsidence due to gas extraction in the nothern part of Netherlands. Land Subsidence, IAHS Publication No 200, pp 99 - 108..

Powell, L. R. and Youbrough, R.E.(1988): Analysis of level survey data over mine subsidence events in Illinois, USA - Abandoned and active mines. Proc. CONFERENCE 1988, Fredericton, New Burnswick, Canada, pp 430 - 439.

Powell, L. R., et. al. (1986): Measurement and analysis of foundation tilt resulting from mine subsidence in Southern Illinois. Proc. 2nd. Work. on Sur. Subsidence due to Underground Mining, Morgantown, WV, USA, pp 142 - 152.

Prasad, R. K. and Rai, R. S. (1981): Some experiences with protection of aerial ropeway over underground workings. Mine Subsidence, Parijat Mudra. Pub., Dhanbad. pp 136 - 142.

Prasad, R.M.(1987): Phenoemnon of sudden subsidence in Jharia and Ranigunj coalfields. IM&EJ, April'87, pp 9 - 17.

Prasad, R. M.(1989): An overview of subsidence and associated problems of Jharia coalfield. MINETECH, v 10, n 1.

Prasad, R.M.(1989): An apparisal of subsidence problems in Jharia coalfield. Land Subsidence, Oxford & IBH Pub., New Delhi. pp 207 - 230.

Prasad, S. K. and Kumar, R.(1989): Assessemnt of subsidence in Indian coal mines. Land Subsidence, Oxford & IBH Pub., New Delhi. pp 469 - 482.

Pratt, W. E. and Johnson, D. W.(1926): Local subsidence of Goose creek field. J. of Geology, v 34, n 1, pp 577 - 590.

Price, D. G., Malkin, A. B. and Knill, J. L.(1969): Foundation of multi-story blocks on the coal measures with special reference to old mine workings. Q. J. Engg. Geol., v 1, pp 271 - 322.

Priest, A. V. and Orchard, R. J. (1957-58): Recent subsidence research in the Nottinghanshire and Derbyshire coalfields. Trans. I. Min. Engr., v 117, p 499.

Prokopovich, N. P.(1984): Classification of land subsidence by origin. Land Subsidence, IAHS Publication No151, pp 281 - 290.

Prokopovich, N. P.(1984a): Origin and treatment of hydrocompaction in the San Joaquin Valley, CA, USA. Land Subsidence, IAHS Publication No 151, pp 537 - 546.

Prokopovich, N. P.(1984b): Economic impact of subsidence on water conveyance in California's San Joaquin Valley, USA. Land Subsidence, IAHS Publication No151, pp 795 - 804.

Prokopovich, N. P.(1991): Detection of aquifer susceptibility to land subsidence. Land Subsidence, IAHS Publication No 200, pp 27 - 34.

Prokopovich, N. P.(1991a): Subsidence and quaternary glaciation. Land Subsidence, IAHS Publication No 200, pp 351 - 358.

Prokopovich, N.P. and Isom, J.(1991): Filled 'subsidence' cracks in San Luis Canal excavations, California, USA. Land Subsidence, IAHS Pub. No 200, pp 379 - 387.

Prokopovich, N.P. and Marriott, M.J.(1983): Cost of subsidence to the central valley project, CA. Bull. of the Assn. Engg. Geologist, v 20, n 3, pp 297 - 315.

Prudic, D. E. and Williamson, A. K. (1984): Evaluation of a technique for simulating a compacting aquifer system in the Central valley of California. Land Subsidence, IAHS Publication No 151, pp 53 - 64.

Pryke, J. F. S.(1954): Eliminating effects of subsidence. Colly. Engg., v 31, p 501.

Qian, J.H.(1985): Consolidation of soils under static and dynamic loading. Proc. 5th Int. Conf. on Numerical Methods in Geomechanics, NAGOYA, Japan., v 1, pp 105 - 110.

Qian, S.Y. et.al.: Computation of Shanghai land subsidence. Chinese J. of Geotech. Engr., v 3.

Qingzhi, Z.(1989): Approach of factors of land subsidence and prevention in Tianjin urban area. Land Subsidence, Oxford & IBH Pub. New Delhi. pp 639 - 658.

Qingzhi, Z. and Xioujun, N.(1984): Analysis of the cause of land subsidence in Tianjin, China. Land Subsidence, IAHS Publication No 151, pp 435 - 444.

Rahman, Md.H. (1996) : Influence of geostructural aspects in land subsidence. Land Subsidence, IAHS Publication, No.234, pp.83-94

Ramnarong, V., Buapeng, S., Shibasaki, N. and Kamata, A. (1995) : Groundwater levels and land subsidence in the Bangkok Metropolitan Area. Land Subsidence, IAHS Publication, No.234, pp.197-206.

Ramnarong, V.(1987): Sub-surface injection of storm water runoff into underground storage of Bangkok. Proc. Int. Symp. on Artificial Recharge of Groundwater, Anaheim, USA.

Ramnarong, V. and Buapeng, S.(1978): Ground Water Level Map of Bangkok. Mitigation of Ground Water Crisis and Land Subsidence in Bengkok Metropolitan Area Project, Dept. of Mineral Resources, Bangkok.

Rankilor, P. R.(1984): The prediction of surface subsidence and strains through thick superficial deposits due to collapse of room-and-pillar mine workings. Mine Workings 84, Proc. Int. Conf. on Construction in Areas of Abandoned Mine Workings, Edinbergh, Soctland.

Rankin, W.J.(1987): Ground movements resulting from urban tunelling, predictions and effects. Proc. Conf. Eng. Group Geological Sci. Engineering Geology of Underground Movements, Nottingham University., pp 13 - 33.

Ranson, B. B., O'Rourke, J. E. and Rey, P. H.(1982): Design and Demonstration of Subsidence Monitoring Systems. Washington DC., Dept. of Energy. Contract No. DE-AC22-80PC30117.

Reddish, D.J.(1984): Study of Ground Strain in Relation to Mining Subsidence. Ph.D. Thesis, Univ. of Nottingham, UK.

Rellensmann, O. and Wagner, E.(1957): The effect on railways of the ground movements due to mining. Proc. European Cong. Gr. Movement., Leeds, pp 74 - 82.

Ren, G., Reddish, D.J. and Whittaker, B.N. (1987): Mining subsidence and desplacement prediction using influence function methods. Min. Sci. Tech., pp 89 - 104.

Rietveld, H.(1984): Land subsidence in the Netherlands. Land Subsidence, IAHS Publication No 151, pp 455 - 465,.

Riley, F.S.(1969): Analysis of borehole extensometer data from central California. Land Subsidence, IAHS Publication, v 89, pp 423 - 431.

Riley, F.S.(1970): Land Surface Tilting Near Wheeler Bridge, Southern San Joaquin Valley, California. US Geological Survey Professional Paper, 497 G. pp G 13 - G 18.

Rimant, A.(1968): Abbau von schachtsieherheitspfeilern. Freib, Forsch.-H, A 448, pp 157 - 173.

Rivera, A., Ledeux, E. and de Marshily, G. (1989): Analysis of compressibility and permeability of Maxico city clay undergoing compaction with a nonlinear quasi-3D model. Land Subsidence, Oxford & IBH Pub. New Delhi. pp 295 - 310.

Rivera, A., Ledoux, E. and de Marsily, G.(1991): Nonlinear modeling of groundwater flow and total subsidencce of the Maxico city aquifer-aquitard System. Land Subsidence, IAHS Publication No 200, pp 45 - 58.

Robotti, F., and Rossini, T.(1984): Analysis of differential settlements on monumental structures by means of the 'DAG' automatic measuring device of levels and inclinations. Land Subsidence, IAHS Publication No 151, pp 347 - 358.

Rodhe, M.W.(1988): Automated data acquisition for engineering and deformation surveys using electronic theodolites. Proc. CONFERENCE 1988, Fredericton,New Burswick,Canada.,pp 440-450.

Roenfedt, M.A. and Holmquist, D.V.(1986): Analytical methods of subsidence prediction. Proc. 1985 Conf. of Coal Mine Subsidence in the Rocky Mountain Region. Colorado Springs, Co. Colorado Geological Survey, Special Publication., v 31. pp 191 - 209.

Rom, H.(1964): A limit angle system. Mitt. Markscheidew, v 71, n 4, pp 197 - 199.

Roscoe, K.H. and Burland,J.B.(1968): On the Generalised Stress Strain Behaviour of Wet Clay. Engineering Plasticity, Cambridge Un. Press, pp 535 - 609.

Roscoe, K.H., Schofield, A.N. and Thuraijrajah, A.(1963): Yieiding of clays in states wetter than critical. Geotechnique, v 13, n 3, pp 211 - 240.

Royal Thai Survey Dept.(1988): Report of Investigation of Land Subsidence in the Bangkok Area from February to July 1988. Royal Thai Survey Department, Bangkok.

Rudra, A.K.(1986): Protection of surface features and structures. Proc. Workshop on Protection of Surface Features and Structures in Mining Areas, Dhanbad.

Russo, P.(1984): Results of recent levellings in the region of Modena. Land Subsidence, AIHS Publication No 151, pp 213 - 224.

Ruth, B.E. and Degner, J.D.(1984): Subsidence and collapse features in karstic coastal plains. Land Subsidence, IAHS Publication No 151, pp 575 - 584..

SANGROM,(1986): The Effect of Underground Mining on Surface. ISRM Symp., South Africa National Group on Rock Mech. (SANGROM), Sandton, South Africa, 155 p.

Safai, N. M. and Pinder, G.(1979): Vertical and horizontal land deformation in a desaturating porous medium. Adv. Water Resour., v 2, pp 19 - 25.

Saini, G. S., Swarup, A. and Dube, A. K. (1989): Monitoring of subsidence for rock mass behaviour around tunnel. Land Subsidence, Oxford & IBH Pub., New Delhi, pp 365 - 374.

Sakurai, S.(1988):Strain distribution around tunnels determined by displacement measurements. Proc. CONFERENCE 1988, Fredericton, New Burnswick, Canada, pp 451 - 461.

Salamon, M. D. G. (1983): Linear models for predicting surface subsidence. Proc. Int. Cong. on Rock Mech. Melborne, pp 107 - 114.

Salamon, M. D. G.(1989): Subsidence prediction using laminated linear model. Proc. 30th. US Symp. on Rock Mech., pp 503 - 510.

Salamon, M. D. G. (1967): A method of designing bord and pillar workings. J. S. Afr. Inst. Min. Metall., v 68, pp 68 - 78.

Salamon, M.D.G.(1968): Elastic moduli of stratified rock mass. Int. J. Rock Mech. & Min. Sci., v 5, pp 519 - 527.

Salamon, M.D.G.(1974): Rock mechanics of underground excavations. Proc. 3rd. Cong. of Int. Soc. for Rock Mech. Denver, Advances in Rock Mech., National Academy of Sci., Washington, 1 (B), pp 951 - 959.

Salamon, M. D. G.(1977): The role of linear models in the estimation of surface ground movements induced by mining tabular deposits. Proc. Conf. Large Ground Movements and Structures, Cardiff, Pentech Press, pp 187 - 208.

Salamon, M.D.G.(1963-64): Elastic analysis of displacements and stresses induced by the mining of seam or reef deposits. J.S. African Inst. Min. Met., pp. 128-149,197-218,468-500., v 64.

Salas, J. A. J.(1979): Two subsidence cases in Spain. Proc. ISRM 4th. Int. Cong. on Rock Mech., Montreux, v 3, pp 163 - 168.

Samanta, S., Sinha, R., et.al.(1986): Subsidence research in Ranigunj coalfield - Ningha Sub-Area. Trans. MGMI, India, v 83, n 2, pp 1 - 10.

Sandoval, J. P. and Bartlatt, S. R.(1991): Land subsidence and fissuring in the Central Arizona Project, Arizona. Land Subsidence, IAHS Publication No 200, pp 249 - 260.

Sar, N. and Malik, A. B.(1992): Depillaring of thick seam in Ranigunj coalfield. Thick Seam Mining, Oxford & IBH Pub., New Delhi. pp 89 - 108.

Sassa, K. et.al.(1980): Lnad-slide triggered by vertical subsidence. Proc. Int. Symp. on Landslide, New Delhi, 1979- 80.

Sato,T.et.al.(1978): Improved model of the Nobi groundwater basin to solve the subsidence problem. Proc. of Speciality Session Geotechnical Engg. and Env. Cont., 9th.ICSMFE, v 2, pp 79 - 81.

Sato, T., Daito, K. and Ueshita, K.(1984): Analysis of land subsidence due to withdrawal of groundwater in the Nobi Plain. Land Subsidence, IAHS Publication No 151, pp 89 - 100.

Saxena, N.C.(1980): Ground movements due to underground extraction of coal seams, vis-a-vis, protection of surface features and structures.

Proc. Adv. Course on St Cont. and Gr. Movement, CMRS, Dhanbad.

Saxena, N. C.(1986): Reclamation and utilisation of subsided land in mining areas. Proc. Collo. on Water Poll. and Land Reclamation in Mining Areas with Special Reference to Jharia Colafield, Dhanbad, pp 37 - 49.

Saxena, N. C.(1986): Monitoring of subsidence movements and interpretation of data. Proc. Workshop on Protection of Surface Features and Structures in Mining Areas, CMRS, Dhanbad.

Saxena, N. C.(1990): Environmental impacts of underground mining. Environmental Management of Mining Operations, Ashish Pub. House, New Delhi, pp 208 - 261.

Saxena, N.C.(1990): Subsidence studies in Jharia coalfield. Proc. Indo-US workshop on Conceptual Environmental Management Plan for Jharia Coalfield, CMPDIL, Ranchi.

Saxena, N. C.(1990): Environmental impacts of underground mining. Proc. Sem. Env. Management of Mining Operations. BHU, Varanasi, pp 258 - 261.

Saxena, N. C.(1991): Subsidence management in Jharia coalfield, India - a concept. Land Subsidence, IAHS Publication No 200, pp 181 - 194.

Saxena, N.C.(1992): Can underground voids in coal mines be used for storage or any other purpose. Proc. Indo-US Workshop on Underground Space Technology, New Delhi, p VI, pp 21 - 23.

Saxena, N.C.(1993): Subsidence problems in Indian Coalfields. J. Min. Met. & Fuels. v.XLI, n 11+12, pp 363 - 383.

Saxena, N. C.(1994): Issues in land use planning in coal mining areas. Proc. 2nd. Nat. Sem. on Mienral & Ecology, Dhanbad, pp 201 - 208.

Saxena, N. C. and Singh, B.(1975): Effect of extraction of a 7-m thick coal seam on Adra-Gomoh railway line in Jharia Coalfield. Proc. IE & MGMI, Meeting, ISM, Dhanbad.

Saxena, N.C. and Singh, B.(1976): Extraction of a thick seam with hydraulic sand stowing underneath a railway line. Proc. 3rd. Int. Symp. Min. Sur., Leoben, Austria.

Saxena, N. C. and Singh, B.(1979): Ground movement research techniques. Proc. Int. Symp. on Mining Research Techniques and Instrumentation, BHU, Varanasi.

Saxena, N.C. and Singh, B.(1980): Investigations into the safety of a railway line against ground movements due to extraction of two thick seams in India. Proc. 21st. US Symp. Rock Mech., Missouri, Dolla.

Saxena, N. C. and Singh, B.(1981): Subsidence and fires in coal seams. Proc. National Sem. on Combating Coal Fires, Dhanbad.

Saxena, N. C. and Singh, B.(1982): Extraction of coal seams underneath surface features and structures in India. Mine Subsidence, Parijat Mudra. Pub. pp 3 - 37.

Saxena, N. C. and Singh, B.(1984): Subsidence research in India. Land Subsidence, IAHS Publication No 151, pp 661 - 668.

Saxena, N. C. and Singh, B. (1985): Subsidence - problems and research in Ranigunj Coalfield. Proc. Sem. on Ranigunj Recontruction - Challenges and Prospects, Dishergarh.

Saxena, N. C. and Singh, B. (1985): Surface subsidence and environmental effects of underground mining. Proc. 3rd. Symp. Rock Mech., Rookee, p III, pp 5 - 13.

Saxena, N. C. and Singh, B.(1986): A status report on problems of subsidence in coal mining areas. J. Min. Met. & Fuels, pp 435 - 445.

Saxena, N. C. and Singh, B. (1986): Subsidence and surface environment. Proc. Symp. Env. Poll. in Dhanbad Coalfields, Dhanbad.

Saxena, N. C. and Singh, B. (1988): Extraction of coal seams underneath railway line at Sudamdih in Jharia Coalfield. Proc. Int. Symp. Underground Engg., New Delhi, pp 389 - 396.

Saxena, N. C. and Singh, B.(1988): Prediction of subsidence movements in coal mining areas in Indian coalfields. Proc. CONFERENCE 1988, Fredericton, Canada, pp 462 - 469.

Saxena, N.C. and Singh, B.(1988): Subsidence in Indian coalfields. Proc. Int. Conf. Gr. Cont., West Virginia, USA.

Saxena, N. C. and Singh, B. (1990): CMRS research report No. MT/PF/13/84, CMRS Dhanbad.

Saxena, N. C. et.al.(1979): Ground subsidence research in India. Proc. IV ISM Aachen, v 4, pp 133 - 148.

Saxena, N. C., Kumar, B. and Singh, B. (1989): Subsidence research in India. Land Subsidence. Oxford & IBH Pub., New Delhi. pp 177 - 192.

Saxena, N. C., Kumar, B., Singh, T. N. and Singh, B. (1993): The 'blind' backfilling of inaccessible coal workings in Indian Coalfields. Proc. 5th. Int. Symp. on Mining with Backfill, Johannesburg, South Africa.

Saxena, N. C., Parti, S. K., Kumar, B., Roychowdhury, S.K. and Singh, B. (1989): Stabilisation of unapproachable underground workings in coal mining areas. Land Subsidence, Oxford & IBH Pub., New Delhi. pp 659 - 670.

Saxena, N. C., Parti, S. K., Kumar, B. and Singh, B.(1988): Blind backfilling of unapproachable workings underneath surface properties. Trans. MGMI, India, v 81, n 2, pp 131 - 148.

Saxena, N. C., Parti, S. K., Kumar, B. and Singh, B.(1988): Stabilisation of unapproachable underground workings in Indian Coalfields - pilot trial. Proc. CONFERENCE 1988, Fredericton, Canada, pp 470 - 475.

Saxena, N. C., Samanta, S., Prasad, S.N., Pandey, B. N., Tiwari, S. and Sinha, R.(1989): Subsidence in Ranigunj coalfield. Land Subsidence, Oxford & IBH Pub., New Delhi. pp 391 - 406.

Saxena, N. C., Singh, R., Samanta, S. and Sinha, R. K.(1992): Extraction of three thick seams underneath a main railway line - A case study. Thick Seam Mining, Oxford & IBH Pub., New Delhi. pp 665 - 676.

Saxena, N. C., Singh, T. N., Singh, K.S. and Kumar, B.(1992): Hydraulic backfilling of unapproachable areas. Proc. National Workshop on Hydraulic Stowing in Indian Mines, Dhanbad, pp 299 - 306.

Saxena, N. C., Sinha, R., Samanta, S. and Singh, B.(1988): Extraction of three thick seams underneath a main railway line and unapproachable workings at Sudamdih in Jharia Coalfield. Proc. CONFERENCE 1988, Fredericton, Canada, pp 476-481.

Saxena, N. C., Tiwary, S., Pandey, B. N., et.al.(1979): Ground subsidence research in India. Proc. 4th. Int. Symp. Mine Sur. Aachen.

Saxena, S.K.(1978): A review of methods used in investigation of subsidence. Proc. Int. Conf. Evaluation and Prediction of Subsidence, Florida.

Schilizzi,P.(1987):Monitoring and Prediction of Surface Movements Above Underground Mines in the Eastern US Coalfields. Ph. D. Dissertion, VPI SU.

Schleier, O. (1956): Vorausberechnung von Bodenbewegungen. Gluckanf, v 2, pp 452 - 508.

Schmidt, B.(1969):Settlements and Ground Movements Associated with Tunnelling in Soil. Ph. D. Thesis, Univ. of Illinios, USA.

Schmidt, B.(1974): Prediction of settlements due to tunnelling in soil : three case histories. RETC, Proc. 2, pp 1179 - 1199.

Schulte, H. F.(1957): The effects of subsidence on strata immediately above a working with different types of packing in level measures. Proc. Europ.Cong.Gr.Movement, pp 188-198.

Schumann, E. H. R. (1983): Controlled Subsidence of Blackhill-Greenside Railway Siding; Part 1. Chamber of Mines of South Africa, Research Report, 8/83.

Schumann, E. H. R. (1986): The monitoring computation and data analysis of surface subsidence. Proc. SANGROM Symp., pp 13 - 23.

Schumann, E.H.R.(1988): The control of subsidence by width-depth ratio and chain pillar size in the presence of competent coal measures. Proc. 7th. Int. Conf. Gr. Cont. in Min., Morgantown, USA, pp 358 - 368.

Schumann, E.H.R.(1989): Subsidence above total extraction panels at South African collieries. Int. St. Cont. Cong. Dusseldorf, pp 134 - 141.

Schumann, H. H., Cripe, L. S. and Laney, R. L.(1984): Land subsidence and earth fissures caused by groundwater depletion in Southern Arizona, USA. Land Subsidence, IAHS Publication No 151, pp 841 - 852.

SchwaRtz, D. and Dubois, S.(1957): Effects of the treatment of the goaf (

Strip packing or caving) on the rock in the immediate vicinity. Proc. Europ. Cong. Gr. Movement, pp 152 - 158.

Scot, R. F.(1979): Subsidence : A review, from "Evaluation and prediction of subsidence" ASCE, pp 1-25.

Selby, A.R.(1987): Surface movements caused by tunnelling in two-layer soil. Proc. Conf. Eng. Group Geological Soc., Engineering Geology of Underground Movements, Nottingham Univ., pp 1 - 12.

Sen, S.K. and Choudhary, S.K.(1982): Pros and cons of extraction of coal seams underneath surface features and structures at Sudamdih project. Mine Subsidence, Parijat Mudra. Pub., Dhanbad, India. pp 38 - 46.

Senevizatne, P.(1987): Design of Partial Extraction System for Mining in New South Wales. Ph. D. Thesis, Univ. of New South Wales, Australia.

Shadbolt, C. H.(1972): Subsidence engineering. Univ. Nottingham Min. Dept. Mag., v 24, pp 80 - 89.

Shadbolt, C. H.(1978): Mining subsidence - historical review and state-of-the-art. Proc. Conf. on Large Ground Movements and structures, Cardiff, Wales, pp 705 - 748.

Shadbolt, C.H. and Mabe, W.J.(1968): Ground Movement in the East Midlands coalfields. Paper presented to RICS (Minerals Section), Basford, Nottingham, UK, 22 p .

Shadbolt, C. H., Whittaker, B. N. and Forrester, D. J.(1973): Recent developments in mining subsidence engineering. Min. Dept. Mag., Univ. of Nottingham, v 35, pp 59 - 74.

Shaham, G. and Levin, I.(1989): Elevation loss in the reclaimed Hula Swamp in Israel during the period 1958-1987.Land Subsidence, Oxford & IBH Pub., New Delhi. pp 323 - 340.

Shalaghinov, N.F.(1963):The limit angles of ground movement. Ugol, v 11, 40 p.

Shankar, K.V.(1987): Subsidence Prediction Using Three Dimensional Numerical Modelling Technique. Ph. D. Thesis, BHU, Varanasi. 69 p.

Shankar, K. V. and Dhar, B. B. (1988): Subsidence Prediction Resulting from Underground Mining - A Numerical Modelling Technique. Proc. Indian Geotech. Conf., Allhabad, pp 319-321.

Shankar, K. V., Dhar, B. B. and Sharma, U. K.(1985): Analysis and prediction of ground movement - a laboratory model. Proc. 6th.Int.Soc. for Mine Surveying, Harrogate, UK., pp 921-927.

Shanker, K. V. and Dhar, B. B.(1989): Subsidence prediction based on numerical modelling technique. Land Subsidence, Oxford & IBH Pub., New Delhi. pp 113 - 120.

Sharp, J. M. (Jr.), Raymond, R. H., Germiat, S. J. and Paone, J. G.(1991): Re-evaluation of the causes of subsidence along the Texas Gulf of Mexico coast and some extrapolations of future trends. Land

Subsidence, IAHS Publication 200, pp 397 - 406.

Shah, S. S. and Husain, M.(1989): The impact of environment on the subsidence of rock masses in the hilly region. Land Subsidence, Oxford & IBH Pub., New Delhi. pp 545 - 554.

Shen, Xia Yu.(1987): Mathematical model and prediction of subsidence in Ning Bo City. Urban Geology of Coastal Lowlands in China, Atlas of Urban Geology, v 3.

Sheorey, P. R. and Singh, B.(1989): Pillar design in partial extraction for subsidence control in coal mining areas. Land Subsidence, Oxford & IBH Pub., New Delhi, pp 599 - 610.

Shimizu, M.(1991): Finite element analysis of land subsidence due to the variation of ground water level. Land Subsidence, IAHS Publication No 200, pp 645 - 654.

Shu-guang, S., Jui-hui, C. and Kai-yao, X.(1984): A mathematical model for the head variation of aquifer and the prediction of land subsidence. Land Subsidence, IAHS Publication No 151, pp 157-167.

Shyam, P. C.(1986): Protection of surface structures against subsidence due to mining ; theory and experience influencing legislation. Proc. Workshop on Protection of Surface Feature and Structures in Mining Areas, Dhanbad.

Sialva-Perez, A. A.(1984): Compaction and subsidence : A critical state approach. Proc. Conf. on Compressibility Phenomenon in Subsidence.

Silva-Perez, A. A.(1984): Analysis of land subsidence using an elastoplastic critical state soil model. Land Subsidence, IAHS Publication No 151, pp 43 - 52.

Sinclair, J.(1963): Ground Movement and Control at Collieries. Sir Issac Pitman and Sons Ltd. London, 349 p.

Sinclair, J.(1950-51): Mining subsidence in South Yorkshire coalfield. Trans. I. Min. E., v 110, 365 p.

Singh, B.(1982): Coal extraction underneath surface features and structures - harmonious mining. Mine Subsidence, Parijat Mudra. Pub., Dhanbad, India, pp 1 - 2.

Singh, B. and Saxena, N. C.(1980): Scientific approach towards conservation of coal resourse in India. Proc. Symp. on Conservation on Mineral Resources, Bhubneshwar,

Singh, K. B., Kumar, P. et.al. (1992): Subsidence behaviour of motour clays. J. Min. Met & Fuels., v XL, n 2, pp 93 - 100.

Singh, K. B., Kumar, P., Kiran, S., Saxena, N.C. and Singh, B.(1989): Subsidence in Kamptee coalfield. Land Subsidence, Oxford & IBH Pub., New Delhi. pp 483 - 496.

Singh, K. B., Samanta, S., Saxena, N. C. and Singh, B.(1990): Subsidence and its environmental impacts at Silewara. Trans. MGMI, India, v 87, n 1, pp 49 - 59.

Singh, K. B., Saxena, N. C. and Jethwa, J. L.(1992): Ground movements due to thick seam mining below an aquifer at Hindustan Lalpeth Colliery. Thick Seam Mining, Oxford & IBH Pub., New Delhi, pp 677 - 686.

Singh, K. B., Saxena, N. C. and Jethwa, J. L.(1992): Subsidence studies due to extraction of a thick seam with hydraulic sand stowing - a case study. IM & EJ, April'92, pp 11 - 16.

Singh, M. M. and Bhattacharaya, S.(1987): Proposed criteria for assessing subsidence damage to renewable lands. Mining Engineering, pp 189 - 194.

Singh, S.K. and Saxena, N.C.(1992): Subsidence related disasters - causes, preventions and precautions. MINTECH, v 13, n 2, pp 19 - 24.

Singh, T.N. and Singh, B.(1989): Strata deformation and subsidence studies for optimal exploitation of coal under water and surface features. Land Subsidence, Oxford & IBH Pub., New Delhi. pp 157 - 175.

Singh, T.N. and Singh, B.(1982): Mining underneath and underground features (Role of model investigations) Mine Subsidence, Parijat Mudra. Pub., Dhanbad. pp 65 - 82.

Singh, T. N., Kushwaha, A., Singh, R. and Singh, R.(1992): Strata behaviour during slicing of thick seam at East Katras colliery. Thick Seam Mining Oxford & IBH Pub., New Delhi, pp 237 - 250.

Singh, T.N. and Singh, B.(1968): Angle of draw in mine subsidence. J. Mines Metals & Fuels, v 16, pp 253 - 258.

Sinha, K.M.(1989): Introduce hydraulic backfilling in US mines for subsidence control. Land Subsidence, Oxford & IBH Pub., New Delhi. pp 633 - 638.

Sinha, P.R.(1989): Subsidence of Bharat Coking Coal Limited - an overview. Land Subsidence, Oxford & IBH Pub., New Delhi. pp 7 - 26.

Sinha, R., Sinha, R. K., Samanta, S., Saxena, N.C. and Singh, B.(1989): Extraction of three thick seams with stowing underneath surface properties at Sudamdih. Land Subsidence, Oxford & IBH Pub., New Delhi, pp 407 - 420.

Siriwardane, H.J.(1985): A numerical procedure for prediction of subsidence caused by longwall mining. Proc. 5th. Int. Conf. Num. Meth. in Geomechs., Nagoya, Japan, pp 1595 - 1602.

Siriwardane, H.J.(1988): Mining Induced Subsidence : Effects on Engineered Structures. Geotechnical Sp. Pub. No.19, American Soc. of Civil Engrs., 209 p.

Siriwardane, H.J. and J. Amanat.(1988): Modelling of subsidence caused by longwall mining using finite element and displacement discontinuity method. Proc. 6th. Int. Conf. Num. Meth. in Geomech., Innsbruk, pp 1901 - 1910.

Siriwardane, H. J.(1985): Numerical modelling of behaviour of overburden rockmass assiciated with longwall mining. Proc. 26th. US Symp. Rock Mech., Rapid City, USA., pp 171 - 177.

Skempton, A. W. and Macdonald, D. H.(1965): Allowable settlements of buildings. Proc. Inst. Civ. Engrs., Pt. III, v 5, p 727.

Snelting, H. and van Bruchem, A.J.(1984): The role of land subsidence and damage to buildings in the selection of suitable sites for groundwater extraction. Land Subsidence, IAHS Publication No 151, pp 767-776.

Snowden, J. O.(1984): Drainage - induced land subsidence in metropolitan New Orleans, Louisiana, USA. Land Subsidence, IAHS Publication No 151, pp 507 - 528.

Sodsri, S.(1978): Geohydrology of the Chao Phraya Basin with Respect of Subsidence. AIT M.Sc. Thesis No. 1296, Bangkok, 154 p.

Sovinc, I., Hass, N. and Ribicic, M.(1983): Prediction and evaluation of subsidence above a thick coal seam. Proc. Int. Symp. Num. Models in Geomechanics, Zurich, pp 814 - 819.

Spencer, L. H.(1960): Subsidence research carried out in the Bestwood area of Nottinghamshire. Min. Engr., v 120, n 3, pp 201 - 210.

Spettmann, J.(1962): Ein Vereinfachtes genares Verfahren der Vorausberechnung von Bodenbewegungen mit Tafelwerten, abzuleiten aus jedev beliebigen Verteilungsfunktion. Mitt. Markscheidew, v 69, pp 1 - 17.

Spierenburg, S.E.J., Calle, E.O.F. and Vonk, Z.C. (1996) : Modelling uncertainty components in measurement of surface level and land subsidence. Land Subsidence, IAHS Publication, No.234, pp.207-214.

Sroka, A.(1988): Selected problems in predicting influence of mining induced ground subsidence and rock deformations. Proc. CONFERENCE 1988, Fredericton, New Burnswick, Canada, pp 482 - 492.

Stample, S.T.(1956):A study of problems encountered in multiple-seam mining in Eastern United States. Bull. of the Virginia Politechnic Institute, v XLIX, n 5, p 64.

Starfield, A. M. and Crouch, S. L.(1973): Elastic analysis of single seam extraction. Proc. 14th. US Symp. Rock Mech., Penn., pp 421 - 439.

Statham, I. and Treharne, G.(1991): Subsidence due to abandoned mining in the South Wales coalfield, UK : causes, mechanisms and environmental risk assessment. Land Subsidence, IAHS Publication No 200, pp 143 - 152.

Statton, C.T., Agnew,J., Naumoff, P., Quittmeyer, R.C. and Johnson D. O.(1984): Ground deformation in the Gulf Cost Region of the US associated with fluid withdrawal from a five kilometer deep reservoir.Land Subsidence, IAHS Publication No 151, pp 187 - 194.

Steed, C., Bawden, W. F., Coode, A. M and Mottached, P. (1985): Subsidence prediction for Saskatchewan Potash Mine. Proc. 26th. US Symp. Rock Mech., Rapid City, pp 163 - 170.

Steinberg, G. and Papo, H. P.(1988): Survey of vertical deformations by GPS and precise leveling. Proc. CONFERENCE 1988, Fredericton, New Burnswick, Canada, pp 493 - 502.

Stephens, J. C., Allen, L. H. and Chen, E.(1984): Organic soil subsidence. Man-Induced Land Subsidence. Rev. Eng. Geology VI, The Geological Soc. of America, pp 107 - 122.

Stoughton, H. W.(1979): Application of geodetic engineering to small scale subsidence studies. Proc. 3rd. Can. Symp. Min. Sur. & Defor. Meas. Sadbury, Ontario, pp 151 - 163.

Strelnikov, K.V. and Samarin, V.P.(1966): The duration of surface subsidence when mining seams at larger depths in the Kizelov district. Gorn davlenie i metodika marksheydersk. Rabot, VNIMI, 56, pp 139 - 144.

Su, D. W. H. and Hasenfus, G. J.(1987): Field measurement of overburden and chain pillar response to longwall mining. Proc. 6th. Int. Conf. Gr. Cont. in Mining, Morgantown, WV, USA, p 156.

Subsidence Research Steering Committee, (1987): Prediction of Subsidence Ground Movement Using Mathematical Modelling Techniques. Research Report, Sydney, 55 p.

Schumann, E.H.R.(1984): Surface subsidence due to the extration of moderately thick coal seams at shallow depth. Proc. Int. Conf. at Univ. of Wales, Instn. Sci. & Tech., Cardiff, pp 248 - 263.

Sugawara, K., Obara, Y., Okamura, H. and Wang, Y.(1985): Pre-calculation of surface subsidence due to coal mining. Proc. 5th. Int. Conf. Num. Methods in Geomech., Nagoya, pp 1603 - 1610.

Sutherland, H.J. and Munson, D.E.(1982): Complementary influence function for predicting subsidence caused by mining. Proc. 23rd. US Symp. on Rock Mech., Berkeley, California, pp 1115 - 1121.

Sutherland, H.J. and Munson, D.E.(1984): Prediction of subsidence using complementary functions. Int. J. Rock Mech., Min. Sci. & Geotech. Abstr., v 21, n 4, pp 195 - 202.

Sutherland, H. J., Heeks, A. A. and Taylor, L.M. (1984): Physical and numerical simulation of subsidence above high extraction coal mines. Proc. 2nd. Conf ISRM, pp 65 - 72.

Symposium Proceedings, (1982): State-of-the-Art of Ground Control in Longwall Mining and Mine Subsidence, Honolulu, Hawaii, ASME, 271 p.

Szefer, G.(1964): The effect of mining on the state of stress and strain of the rock mass according to the theory of elasticity. Zeszyty Problemowe Gornictwa, v 1, pp 3 - 54.

Szostak-Chrzanwski, A.(1988): An iterative modelling of ground subsidence using non-linear elastic finite element analysis. Proc.

CONFERENCE 1988, Fredericton, New Burnswick, Canada, pp 524 - 535.

Tandanand, S. and Powell, L. R.(1982): Assessment of Subsidence Data from the North Appalachian Basin for Subsidence Prediction. USBM, RI 8630.

Tandanand, S. and Powell, L.R.(1984): Influence of lithology on longwall mining subsidence. Min. Engg., v 36, pp 1666-1671.

Terzaghi, R.D.(1969): Brine field subsidence at Windsor, Ontario. Proc. 3rd. Symp. on Salt, Cleaveland, pp 298 - 307.

Teskey, W. F. and Proter, T. R.(1988): An integrated method for monitoring the deformation behaviour of engineering structures. Proc. CONFERENCE 1988, Fredericton, New Burnswick, Canada, pp 536 - 547.

Testa, S.M.(1991): Elevation changes associated with groundwater withdrawl and re-injection in the Wilmington area, Los Angeles Coastal Plain, California. Land Subsidence, IAHS Publication No 200, pp 485 - 502.

Tie-Zhu, Liu. and Jie-Jia, Chen.(1989): Quantization theory study of land subsidence effect appreciation in Shanghai. Land Subsidence, Oxford & IBH Pub., New Delhi, pp 103 - 112.

Tiwary, S., Samanta, S., Saxena, N. C. and Singh, B.(1989): Longwall caving below surface properties at Moonidih project. Land Subsidence, Oxford & IBH Pub., New Delhi. pp 455 - 468.

Trahan, D. B.(1982): Monitoring local subsidence in areas of potential geopressured fluid withdrawl, Southwestern Louisiana. Trans. Gulf Coast Assoc. Geol. Soc., v 32, 231 - 236.

Trent, B. C. and Langland, R. T.(1983): Subsidence modelling for underground coal gasification. In Situ, v 7, n 1, pp 53 - 85.

Trevits, M. A. and King, R. L.(1986): Overview of the Bureau of Mines Subsidence Research Program. 88th. Annual Gn. Meeting of CIM, Montreal, 14 p.

Triplett, T.L. and Yurchak, D.W.(1991): Inclusion of an intensity function for subsidence prediction with the influence function approach. Land Subsidence, IAHS Publication No 200, pp 153 - 162.

Tsurlavic,Y. and Denekamp, S.(1980): A boundary element method for analysis of subsidence associted with longwall mining. Proc. 2nd. Int. Conf. at Univ. of Wales, Instt. Sci. & Tech., Cardiff, pp 65 - 74.

Tsurlavic, Y. and Fainstein, G.(1984): Geometry of subsidence associated with longwall mining. Proc. Int. Conf. at Univ. of Wales, Instt. Sci. & Tech., Cardiff, pp 65 - 74.

Ueshita, K. and Daito, K.(1987):Land subsidence in Japan and a case study on the Nobi Plain. Proc. Int. Symp. on Geotech. Aspects of Mass & Material Transportation, A.A.Balkema., pp 439 - 460.

Ueshita, K. and Sato, T.(1980): Study on the groundwater level preventing land subsidence of the Nobi Plain. Trans. JSCE, v 12, pp 188 - 189.

Ueshita, K. and Sato, T.(1981): Study on subsidence of the Nobi Plain. Proc. 10th. Conf. on Soil Mech and Foundation Engg., v 2, pp 387 - 390.

Ueshita, K. et.al.(1977): Modelling of the Nobi groundwater basin to solve subsidence problem. Proc. of Speciality Session on Geotechnical and Env. Conf., 9th. ICSMFE, v 1, pp 465 - 480.

Urbanik, W. and Osborne,V.T.(1989): Surface subsidence prediction above extractions in stratified deposits - a computer model. Proc. Computer Syst. in Australian Min. Industry, Univ. of Woll., NSW, pp 117 - 125.

VNIMI.(1958):The Movement of Rock Masses and of the Surface in the Main Coalfields of the USSR. Gn. Instt. Mine Sur., Leningrad, 250 p.

Van Bruchem, A. J.(1984): Geohydrological aspects of the reclamation of the Markerwaard Podler, The Netherlands. geotechnical investigations. Land Subsidence, IAHS Publication No 151, pp 899-906.

Van Opstal, G. M.C.(1974): The effect of base-rock rigidity on subsidence due to reservoir compaction. Adv. in Rock Mech., US Nat. Ac. of Sci., v 23, pp 1101 - 1111.

Van der Merwe, J.N.(1986): Design methods to arrive at the optimal placing and mining of inter panel pillar to alleviate their effects on surface. Proc. SANGORM 1986, Symp. pp 133 - 144.

Van der Merwe, J.N.(1986): Analysis of surface subsidence over a longwall panel at 50m below surface. Proc. SANGORM 1986, Symp., pp 145 - 150.

Vandale, A.C.(1967): Subsidence : a real or imaginary problems. Min. Engg., v 19, n 10/9, pp 86 - 88.

Vanicek, P. and Kwimbere, M. D. J.(1988): Displacements versus strain. Proc. CONFERENCE 1988, Fredericton, New Burnswick, Canada, pp 557 - 562.

Vartanyan, G. S. and Kulikov, G. V. (1984): Hydrogeodeformation field in studying processes of land subsidence. Land Subsidence, IAHS Publication No 151, pp 101 - 106.

Veith, D.L.(1978): Mined Land Subsidence Impacts on Farmland with Potential Application to Illinois, a Literature Review. USBM, IC 9124, 16 p.

Verhaegh, T.(1990): 'CFD' a computer program for the calculation of compaction from subsidence. Caracas, pp A1 - A2.

Viergever, M.A.(1991): Pore water pressure and subsidence in long term observations. Land Subsidence, IAHS Publication No 200, pp 575 - 584.

Villaneuva, C.C.(1984): Numerical simulation of subsidence due to pumping with hysteresis effect included. Land Subsidence, IAHS Publication No 151, pp 79 - 88.

Virginia Politechnic Inst. & State Univ. (1987): Prediction of Ground Movements due to Underground Mining in the Eastern United States Coalfields, Final Report, Office of Surface Mining, Reclamation and Enforcement, v 2.

Voigt, B. and Pariseau, W.(1968): The Nature of Prediction in Subsidence Engineering. ASCE Meeting, Pittsburgh, Preprint no 762, 42 p.

Voigt, B. and Pariseau, W.(1970): State of prediction art in subsidence engineering. J.Soil Mech., n 96(SM2), pp 721-750.

Voigt, B. and Dahl, H. D.(1969): A post-yield phenomenological approach to mine subsidence. Proc. Int. Symp. Mine Surveying, Parag, v III/3.

Vos, G., Claessen, F.A.M. and Van Ommen, J. H.G.(1984): Geohydrological compensatory measures to prevent land subsidence as a result of the reclamation of the Markerwaard Podler in the Netherlands. Land Subsidence, IAHS Publication No 151, pp 915 - 924.

Wall, J. A. de (1986): On the Rare Type Compaction Behaviour of Sandstone Reservoir Rock. Thesis Techn. Univ. Delft, 165 p.

Wade, C. V. and Canroy, P. J.(1980): Rock mechanics study of a longwall panel. Min. Engr., v 32, n 12, pp 1728 - 1734.

Wade, C. V and Canroy, P.J.(1977): Rock Mechanics Study of a Longwall Panel. Sco. Min. Engrs. Fall Meeting, St. Louis, Mo, Pre-print No. 77-1-391.

Wagener, F. Von, M. et.al.(1989): Surface subsidence in Southern Africa. Proc. XII ISSMFF Conf., Rio de Janerio.

Walker, J. S. and LaScola, J. C.(1988): Foundation response to subsidence induced ground movements - A case study. Proc. CONFERENCE 1988, Fredericton, New Burnswick, Canada, pp 563 - 572.

Walters, R. F.(1977): Land Subsidence in Central Kanas Releted to Salt Dissolution. Kanas Geological Survey Bulletin. No. 214, 82 p.

Walton, G. and Cobb, A. E.(1984): Mining Subsidence. Ground Movements & Their Effects on Structures, Surrey Univ. Press, pp 216 - 242.

Wanless, B.A. and Jones, H.R.(1988): Nortech experience with GPS for precise engineering and subsidence applications. Proc. CONFERENCE 1988, Fredericton, New Burnswick, Canada, pp 573 - 579.

Wardell, L.(1959): The problems of analysing and interpreting obsreved ground movements. Coll. Engg., v 36, n 12, pp 529 - 538.

Wardell, K. and Wood, J. C.(1966): Ground instability problems arising from the presence of old, shallow mine workings. Proc. Midlands Soil Mech. and Foundation Eng. Soc., v 7, n 36, pp 5 - 30.

Wardell, K.(1950): A comparison between British and German experiences of mining subsidence. Trans. Int. Min. Sur., v 30, 51 p.

Wardell, K.(1952):The surveying observations required for the determination of ground movements caused by mining. Trans. Int.

Min. Sur., v 32, 12 p.

Wardell, K.(1953): Some observations on the relationship between fire and mining subsidence. Trans. Ist. Min. Engrs., v 113, pp 471 - 483.

Wardell, K.(1954): Mining subsidence.Trans. R.I.C.S., v 86, 53 p.

Wardell, K.(1957): The minimisation of surface damage. Coll. Engg., v 34, p 361.

Wardell, K.(1957): The minimisation of surface damage by special arrangement of underground workings. Proc. Euorp. Cong. Gr. Movement, pp 13 - 19.

Wardell, K.(1966): Problems posed by past and future mineral workings in a development area. The Chartered Surveyor, Mineral Workings, Leeds, pp 184 - 188.

Wardell, K.(1969): Ground subsidence control. Mining Cong. J., v 55, n 1, pp 36 - 42.

Wardell, K.(1975):Mining Under Tidal Waters. A report prepared for the Minister for Mines and Power (NSW Dept. of Mineral Resources, Sydney.).

Wardell, K.(1953-54):Some observations on the relationship between time and mining subsidence. Trans. Inst. Min. Eng., v 113, pp 471 - 483.

Wardell, K. and Webster, N.E.(1957): Some surface observations and their relationship to movement underground. Proc. Europ. Cong. Ground Movement, Leeds., pp 141 - 148.

Wardell, K. and Webster, N. E.(1957a): Surface observations and strata movement underground. Coll. Engg., v 34, p 329.

Wardell. K. and Eynon, P.(1968): Structural concepts of strata control and mine design. The Mining Engrs, v 127, Aug.'68, pp 633 - 645.

Webb, B.(1982): A Study of Longwall Subsidence in the Appalachian Coalfield. M.S. Thesis, VPI & SU.

Webester, N. E.(1951-52): Strata control and influence on underground and surface damage. Trans. Inst. Min. Engr., v 111, pp 445 - 462.

Weir, A.M.(1966):Subsidence- the interpretation of travelling strain observations. Colliery Guardian, v 212, pp 576 - 577.

Whetton, J. T.(1957): A general survey of the ground movement problem. Coll. Engg., v 34, p 153.

Whetton, J. T. and King, H. J.(1961): The time factor in mining subsidence. Proc. Int. Symp. Min. Research, Rolla, Missouri, v 1, pp 1 - 29.

Whitelock, G.C.H.(1920): Subsidence due to coal mining. Trans. I. Min. Surveyers.

Whittaker, B. N. and Redish, D. J.(1984): Mining subsidence in longwall mining with special reference to the prediction of surface strain. Proc. 2nd. Int. Conf. on Stability of the underground Mining, NewYork, pp 576 - 588.

Whittaker, B. N.(1983): Chain pillar design considerations with reference to longwall mining. Mining Dept. Magazine, Univ. of Nottingham, pp 65 - 76.

Whittaker, B. N. and Aston, T. R. C. (1958): Under-sea longwall mining subsidence with special reference to geological and water occurrence criteria with North-East of England coalfield. Min. Sci. & Tech., v 2, pp 105 - 130.

Whittaker, B.N. and Breeds, C.D.(1977b): The influence of surface geology on the character of mining subsidence. The geotechnics of structurally complex formations. Geotechnica - Italia, Millan, v 1, pp 459 - 468.

Whittaker, B. N. and Frith, R.C.(1987): Aspects of chain pillar design in relation to longwall mining. Proc. 6th. Int. Conf. Gr. Cont. in Mines, Morgantown, WV, USA, pp 172 - 182.

Whittaker, B. N. and Pasamchmetogly, A.G.(1981): Ground tilt in relation to subsidence in longwall mining. Int. J.Rock Mech.& Min.Sci.& Geomech. Abstr., v 18, pp 321 - 329.

Whittaker, B.N. and Pye, J.H.(1977a): Ground movements associated with the near surface construction operations of a mine drift in coal measure strata. Int. J. Rock Mech. Min. Sci. & Gr Abstr., v 14, pp 67 - 75.

Whittaker, B.N. and Pye, J.H.(1977b): Design and layout aspects of longwall methods of coal mining. Proc. US 16th. Symp. Rock Mech., Minneapolis, pp 303 - 314.

Whittaker, B. N. and Redish, D. J.(1989): Subsidence : Occurence, Prediction and Control. Elsevier, Amesterdam, 508 p.

Whittaker, B. N. and Singh, R. N.(1979): Design and stability of pillars in longwall mining. The Min. Engr., v 214, pp 59-70.

Whittaker, B.N. and Smith, S.F.(1987): Stability and operational aspects of room and pillar mining in the UK sedimentary iron-ore deposits. Underground Mining Methods and Technology, Elsevier, pp 393 - 402.

Whittaker, B. N., Reddish, D. J. and Fitxpatrick, D. J.(1985a): Calculation by computer program of mining subsidence ground strain pattern due to multiple longwall extractions. J. Min. Sci. & Tech., v 3, pp 21 - 33.

Whittaker, B. N., Reddish, D. J. and Fitzpatrick, D. J.(1985b): Ground fracture due to longwall mining subsidence. Proc. I.M.W.A. Second Int. Cong. on Mine Water, Granada, Spain, pp 1057 - 1072.

Wiborg, R. and Jewhurst, J.(1986): Ekofish subsidence detailed and solution assessed. Oil and Gas J. (Technology), pp 47 - 51.

Wildanger, E. D. et.al. (1980): Sink-hole Type Subsidence Over Abandoned Coal Mines in St. David, Illinois. Abandoned Mined Lands Reclamation Council, Springfield, Illinois.

Williamson, A.K. and Prudic, D.E. (1984): Simulation of flow and compaction in the regional aquifer system of the central valley of California, USA. Land Subsidence, IAHS Publication 151, pp 271-280.

Williams, J.H. and Vineyond, J.D.(1976): Geological indicator of catestropic collapse of karst terrane in Missouri. National Academy of Sci. Transpotation Research Record 612, pp 31-37.

Wilson, A. H.(1972): Research into the determination of pillar size. Mining Engr., v 131, pp 409 - 416.

Wilson, A. H.(1972): An hypothesis concernning pillar stability. Mining Engr.

Winstanley, A.(1938): Strata movements. Trans. Inst. Min. Sur., v 19, pt.I.

Williamson, A.S.(1979): Volumetric subsidence - compaction ratios for a horizontal, Stratified elasic half space. Adv. in Rock Mech., v 2 B, pp 1112 - 1120. US Nat.Ac.of Sci.

Wilschut, D. (1995) : Long term settlement in Rotterdam. Land Subsidence, IAHS Publication, No.234, pp.215-220.

Wolf, R.G.(1970): Relationship between horizontal strain near and reverse water level fluctuation. Water Resources Research, v 6, n 6, pp 1721 - 1728.

Wu, W. and Haycocks,C.(1986): Statistical analysis of problems in close proximity multi-seam mines. Proc. 27th. US Symp. Rock Mech., Tuscaloosa, AL., pp 317 - 323.

Wu, W., Haycock, C. and Zhou,Y.(1987): Applications in design for close proximity multi-seam mining. Proc. 6th. Int. Conf. Gr. Cont. in Mining, Morgantown, WV, USA, pp 97 - 105.

Wyllie, P. J.(1971): The Dynamic Earth. Textbook in Geosciences, John Wiley and Sons, Inc., 416 p.

Xu, D. M., Yong, R. N. and Mohamed, A. M. O. (1991): Time delay effect of subsidence. Land Subsidence, IAHS Publication No 200, pp 635 - 645.

Yamamoto, S. and Kobayashi, A.(1984): Groundwater resources in Japan with special reference to its use and conservation. Land Subsidence, IAHS Publication No151, pp 381 - 391.

Yao, X. L., Whittaker, B. N. and Reddish, D.J.(1991): Influence of overburden mass behavioural properties on subsidence limit characteristics. Min. Sci. and Tech., v 13, 167 - 173.

Yashida, S.(1983): One Dimensional Simulation of Land Subsidence During the Period of Piezometric Head Rise in Aquifers. Ann. Rep. Of the Research Institute of Hazards, Niigata Univ., n 5.

Yokel, F. Y., Salamone, L. A. and Chugh, R. M.(1981): Construction of Housing in Mine Subsidence Areas. Report No. NBSIR 81- 2215, Washington, DC: US Dept. of Commerce, National Bureau of Standards.

Yong, R.N. and Nutalaya, P. (1988): Impact of Quarternary Sediments on Urban Development and Land Use of the Central Plain of Thailand. Final Rep. Submitted to the Int. Development Research Centre (IDRC), Ottawa, 242 p.

Yong, R. N., Nutalaya, P., Mohamed, A.M.O. and Xu, D. M. (1991): Land subsidence and flooding in Bankkok. Land Subsidence, IAHS Publication No 200, pp 407 - 417.

Yoshida, S.(1982): A consideration of mechanism of land subsidence during the period of piezometric head rise in aquifers. Ann. Rep. of the Research Institute of Hazards, Niigata Univ., n 4, pp 99 - 106.

Yoshida, S. and Aoki, S.(1984): Mechanics and one-dimensional simulation of land subsidence during the period of recovery in aquifers. Land Subsidence, IAHS Publication No 151, pp 147 - 156.

Young, C. M.(1971): Precentage of extraction of bituminous coal with special reference to Illinois conditions. Engineering Experimental Station Bull. No.100. Urbana: Univ. of Illinois.

Young, L. E. and Stock, H. H.(1916): Subsidence resulting from mining. Univ. of Illinois, Engineering Experiment Station, Bull. No. 91, 204 p.

Yu-sun, Chen.(1984): Over recharge exploration of ground water and land subsidence in Taiyuan city, Shanxi Province, China. Land Subsidence, IAHS Publication No 151, pp 415 - 424.

Yudhbir, (1984): Mechanisms of collapse controlled subsidence. Land Subsidence, IAHS Publication No 151, pp 467 - 478.

Zamare, M. M., Ahern, J. L. and Najjar, Y. M.(1989): Stability analysis and characterisation of ground subsidence of abandoned lead-zinc mines in North Oklahoma. Proc. 30th. US Symp. Rock Mech., pp 707 - 714.

Zeevaert, L.(1963): Foundation problems related to ground surface subsidence in Mexico city. Field testing of soils. ASTM Special Publication No. 332, pp 57 - 66.

Zeevaert, L.(1973): Foundation Engineering for Difficult Sub-soil Conditions. Van Nostzand Reinhold Company, New York.

Zenc, M.(1969): Comparison of Bals' and Knothe's method of calculating surface movements due to underground mining. Int. J. Rock Mech. Min. Sc., v 6, n 2, pp 159 - 190.

Zhong, W. and Zhu, R.(1979): Calcution of surface movement and deformation resulted from mining steeply dipping seam. Proc. IV ISM, Aachen, Paper No. 4.38.

Zienkiewicz, O.C.(1979): The Finite Element Method. Tata Mc Graw Hill, India, 787 p.

Zilkovski, D.B. and Kok, J.J. (1995) : Estimation of subsidence in New Orleans and vicinity as indicated by precise relevellings. Land Subsidence, IAHS Publication No.234, pp.221-234.

